DESIGN FOR A CENTURY

South British Chambers.
Hereford Street.
Christchurch. March 2nd 1887

Dear Sir

I forward herewith a sketch of a scheme for the establishment of a school of engineering, which I think you will find comprehensive and which can be carried out gradually without much initial expense beyond the establishment of a chair of engineering.

I have sent you my rough draft in order to give you as much leisure as possible for its perusal before the afternoon meeting.

I remain dear sir
yours faithfully
E Dobson

F. de C. Malet Esq

Design for a Century

A HISTORY OF THE SCHOOL OF ENGINEERING,
UNIVERSITY OF CANTERBURY 1887–1987

Diana Neutze & Eric Beardsley

CHRISTCHURCH
UNIVERSITY OF CANTERBURY
1987

University of Canterbury. Publication No. 37

ISBN 0900-392-37-1

DESIGNED AND PRINTED AT
THE CAXTON PRESS, CHRISTCHURCH, NEW ZEALAND

Contents

Illustrations

Preface

Diana Neutze was appointed to write the centennial history of the School of Engineering in 1983, but subsequent ill health compelled her to acknowledge that she would be unable to complete the commission alone. Accordingly the School's Centennial Organising Committee agreed at the end of 1985 that Dr Neutze should write the history of the School's foundation and growth on the University's city site — a period she had already researched — and that the Information Officer, Eric Beardsley, should cover the transformation of the School since the decision in 1949 to move the whole University to Ilam.

The plan demanded rapid research and writing if the book was to be published in time for the centennial celebrations in May 1987, and there was little opportunity for close collaboration between the authors as the work progressed. Accordingly, readers will be confronted with a book in three distinct parts dealing with three distinct periods in two different styles.

But both authors have much in common. They determined that while their book might be of prime interest to members of the School — and there have been more than 6000 graduates — it would also attract those who simply enjoy reading history. They also determined to set the School in the wider contexts of College and University, city, province and country to give a broader perspective to its life and work.

The authors have presented as faithful and accurate a record of the School's first 100 years as was possible in the time available. If their selection and interpretations are challenged, if errors of fact have escaped their vigilance and if some worthy contributors to the School's life have been overlooked, they can only reply that what has been done has been done honestly within the constraints of time available to them. The book is not an "official" history by any means; it is simply how two interested writers saw the development of the University's major special school.

The authors gratefully acknowledge the support of the Editorial Subcommittee of the Centennial Organising Committee, comprising Professor David Elms (chairman), Mr John Pollard, Dr George Mullenger and, until his untimely death, Professor H.J. Hopkins; and of the University Publications Committee, particularly its chairman, Mr R.W. Hlavac, and its editor, Mr A.N. Brooks.

They also wish to thank many members of the staff, past and present, for their assistance and the following in particular: Dr James Watson, Dr T.G. Emery, Professor H.N. Parton, Mrs Shirley Shea, Miss Jan Kotlowski, Miss Angela Saggins, Mrs Margaret Sutton, Miss J. Kane, Mr F. Whiteling, Mrs Jillian Scandrett, Miss Carolyn Reid and Mr P.M. Johns.

They are also indebted to the University Library, the Engineering School Library, the *Star* and the *Press*, Christchurch, the Archives of Canterbury Library, and the Canterbury Public Library. The use of photographs from the Canterbury Museum is gratefully acknowledged.

Also acknowledged with gratitude are financial contributions towards the writing of the book from the following firms:

Alan Reay Consultants (Christchurch)
Beca Carter Hollings and Ferner Ltd (Wellington)
Esso Australia Ltd (Sale, Sydney)
Evans, Douglas (Christchurch)
Hadley and Robinson Ltd (Dunedin)
Halliday, O'Loughlin and Taylor (Christchurch)
Holmes Wood Poole and Johnstone (Christchurch)
KRTA Ltd (Wellington)
Mobil Oil Ltd (Wellington)
Morrison, Cooper and Partners (Wellington)
Royds Garden Ltd (Christchurch)
R. W. Morris and Associates, Consulting Engineers Ltd (Christchurch)
Smith and Leuchars Ltd (Wellington)
Tyndall and Hanham (Christchurch)

Abbreviations

The authors have used two kinds of notes: footnotes, which appear at the foot of the appropriate page, to amplify a point or to provide brief biographical information; and source notes, indicated by superior numbers in the text, to provide the source and bibliographical references on which the text is based. The references may be found at the end of the book. The meaning of most abbreviations should be clear from the context, but the following are spelt out to avoid any ambiguities:

AR	Annual Report of Canterbury (University) College
AU(C)	Auckland University College
BG	Board of Governors, Canterbury College
BM	Minutes of the Board of Governors, Canterbury College
CU	University of Canterbury
C(U)C	Canterbury (University) College
CM	Minutes of the Council, Canterbury University College and the University of Canterbury.
educ	Educated
LI	Letters inward (to the College)
LO	Letters outward (from the College)
LT	*Lyttelton Times*, Christchurch
P	*The Press*, Christchurch
S	*Star-Sun* and *Christchurch Star*, Christchurch
PBM	Minutes of the Professorial Board, University of Canterbury
PVC	Pro-Vice-Chancellor
UNZ	University of New Zealand
UR	University of Canterbury Registry
VC	Vice-Chancellor
VUC	Victoria University College

Part I: Engineering Potentate

CHAPTER ONE

Introduction: 1849-1887

On 26 July 1887, Edward Dobson, the former provincial engineer, gave the inaugural address of the School of Engineering. It did not attract a large audience. But then, it was a School only in name, consisting of no special building, a smattering of equipment collected over the years and housed in the Department of Technical Science at the Museum, apprentice students drawn from the various industries and trades of Christchurch, and two part-time lecturers, Edward Dobson* himself and Robert Julian Scott,** General Manager of the Addington Workshops. Such a humble beginning was seen to be perfectly appropriate, requiring, as it did, "a comparatively nominal expenditure."[1] In a *Press* interview of 22 February 1887, the Professor of Geology, F.W.Hutton, with no premonition that his assertions were shortly to be disproved, had announced: "We cannot go in for an elaborate school here with buildings and classrooms of its own, and students entirely of its own, but luckily these things are not needed." He was perfectly confident that "with such lectures (as those given by Mr Scott of the Addington Workshops, at the School of Art, on mechanical engineering) great practical results

---

*Dobson, Edward, 1816-1908; educ London University and under an architect cum surveyor; worked for John Rastrick's rlwy engineering firm; arr ChCh 1850. As Canterbury's Provincial Engineer he spent 14 years designing, supervising and building the province's principal engineering works including the Lyttelton tunnel. Associated with civil engineering work in the South Island and Australia; m ICE, 1881, received Telford medal for paper on engineering works in Canty; author *Pioneer Engineering* and *Public Works of Canterbury*; he and family commemorated in IPENZ Dobson lectures.

**Scott, Robert Julian, 1861-1930; educ Schl of Mines, London; apprentice rly engineer in England; chief draughtsman of NZ Rlys; manager Addington Rly Wkshops; outstanding designer of rly locos and rolling stock; part-time lect mech eng CC, 1887; lect-in-charge Schl Eng and Tech Sci CC, 1889; prof-in-charge Schl Eng CC, 1894-1923; m Sen UNZ, 1902-23; m of cmsns on NZ rlys. Outside engineering, his passion was yachting. He was a skilled marine designer and keen sailor.

may be obtained at a small outlay."[2] He was not the only one to suffer from the misconception that it was possible to get a School of Engineering on the cheap, nor to misread the character of Mr Scott of the Addington Workshops, thinking him one who would readily co-operate with such a scheme.

While 26 July 1887 celebrated the beginning of the School, it also celebrated the culmination of nearly 40 years of plans, proposals and submissions, political and educational concern and publicity, all directed towards the subject of technical education, its importance and place in a new colony. That Canterbury needed engineers was never in doubt; communication, transport, water, sewerage, such amenities were basic. As well, the very landscape posed fundamental engineering problems. Bishop Selwyn, on his arrival in Lyttelton in January 1851 could complain that "Money enough has been spent but all in civil engineering . . . and not one sixpence of expenditure in any form for the glory of God or the comfort of the clergy."[3] The early settlers, however, were less grudging. After all, they had to struggle up the "veritable mountain"[4] of the Bridle Path on foot, carrying as many of their household goods as they could manage. Uncompromising as the hills undoubtedly were, there were other more effective and dangerous "adversaries"[5] to come: the unpredictable Canterbury rivers, the crossing of which could easily spell "sudden death"[6] and the great mountain chain which separated Canterbury from the wealth of the West Coast gold mines. There was no doubt but that imaginative and adventurous engineers were sorely needed.

Such a need had been recognised in the plans for the new colony even before the early settlers had entrusted their lives to the First Four Ships. At a public meeting in Ipswich on 30 May 1849, the Bishop of Norwich outlined the Canterbury Association's plans for a college — Christ's College — which should have an upper school, an embryonic University College, embracing "four divisions: theology, classical, mathematical and civil engineering, and agricultural."[7] If his description of the civil engineering division — civil engineering at that time being the general name for all engineering that was not military — leant very heavily towards practical rather than theoretical instruction, this was partly a pragmatic assessment of the needs of the burgeoning colony. Also, at that period, there were very few models of the University teaching of engineering that the Bishop could draw on. The first British engineering chair had been established at the University of Glasgow in 1840, but "though engineering classes were offered, no degree was awarded in science or engineering *per se* until the B.Sc. degree was established in 1872"[8] and then only as a result of spirited student protest.

This dream of the upper school of Christ's College was to be realised ultimately in the establishment of the Canterbury Collegiate Union which, in its turn, was transformed into Canterbury College. When the College was established in 1873, by means of a co-operative gesture between the Museum and Christ's College, there was created a Board of Governors, initially 23 men, serving voluntarily, who controlled all the money set aside for education in Canterbury. One of the farsighted endowments of the provincial government, aware of the fast-approaching disbandment of provincial governments in favour of one centralised government and very loath for that central government to get its hands on Canterbury's wealth, was to pass ordinances reserving vast areas of land for the purposes of higher education in the province and to provide, in a separate endowment, for the Museum, Public Library and a School of Technical Science. "The Board of Governors to which these resources were committed had three main functions: to manage an estate, to maintain a number of institutions — [which were to include Boys' High School, Girls' High School, the Canterbury Museum, the Public Library, Canterbury College, the School of Art and the School of Engineering] — and control their several staffs . . . The Board was, in its first two decades and more, the provincial authority in most fields of education above the primary level."[9] There is no need to repeat the story of exactly how Canterbury College was founded. Despite the original plans to offer courses in engineering, when the first enterprising students assembled in 1873, engineering was not included in the curriculum. However, this omission was regarded as purely temporary. H.J.Tancred, the leading figure in the university politics of the 1870s, conferred the opening address of Canterbury College. He indicated his regret that "We have as yet no schools for the acquisition of purely professional knowledge. I hope that no length of time will elapse before we are in a position to establish these also."[10] And at the opening address of the 1879 session of Canterbury College, the Mathematics Professor, C.H.H.Cook,* one of the triumvirate of foundation professors, was to

*Cook, Charles Henry Herbert, 1844-1910; educ Univ of Melb and St John's Coll, Cambridge, of which he became a Fellow; studied mathematics and law; MA (Cantab), 1872; appointed first prof of maths and nat philos at CC, 1874; m Royal Cmsn on Higher Educ, 1878-82; m Sen UNZ, 1884-1908; superintendent Schl of Engineering, 1887-9; prominent in ChCh music; Fellow of Christ's Coll, 1891-1908; first professor emeritus CC, 1909.

declare: "It can hardly be supposed that such classes of men as lawyers and civil engineers will be long left to pick up the substratum of their professional education as best they may".[11]

In 1876, there was a distinct possibility that a School of Mines would be established at Canterbury. The House of Representatives resolved:

> That having in view the vast undeveloped mineral wealth of the Colony, and the necessity of obtaining the fullest possible information concerning the metals and minerals with which the country abounds, the House is of opinion that it is advisable to establish a School of Mines.[12]

The Government requested both Canterbury College and Otago University to submit proposals for such a School. This opportunity was greeted with eagerness by two of the foremost scientists of the province — Sir Julius von Haast* and Professor Bickerton** — both of whom were constantly agitating for the need to provide education in technical science. Von Haast had established the Philosophic Institute in 1862, and both he and the science professor were constantly promoting the cause of technical science by means of popular lectures. They conferred on the question of a School of Mines, and decided it would be expedient and possible with very little outlay. Canterbury enthusiasm waned, however, and the coveted prize went to Otago, not before von Haast had expended a great deal of time and energy gathering a collection of mining models, many coming from his friend, the Austrian geologist, Ferdinand von Hochstetter. These models were to become the major equipment of a Department of Technical Science at the Museum, which was officially opened in 1882; von Haast described them as a "valuable collection . . . for the teaching of mining and metallurgy."[13] But popular lectures and a display of mining models were a totally unsatisfactory substitute for an actual School of Engineering, and indeed many people

---

*von Haast, Sir J.F.Julius, 1822-87; geologist, explorer and museum director;. Educ Univ of Bonn; arr NZ 1858; geological expeditions in Auckland and Nelson, 1859-60; Canty Prov Geologist, 1861-68; first director of Canty Museum, 1869-87; Austrian knighthood, 1875; first prof of geology, Canty Coll, 1876-87; m Sen UNZ, 1880-7.

**Bickerton, Alexander William, 1842–1929; educ Royal School of Mines, where he had outstanding record, 1870–3; FCS; organised working men's science classes, Chelsea; taught at Harley Institute and Winchester College; first professor (in chemistry, and later physics) CC, 1874; his teaching the subject of two enquiries by BG, 1894-5 and 1899-1902; dismissed, 1902; spent rest of his life advocating his "partial impact" theory (first formulated 1878); professor emeritus CC, 1928.

believed that the Museum was already appropriating public moneys which were required for the Public Library. Certainly, von Haast, a dynamic and energetic man, had a way of inveigling money out of the Board of Governors where a less charismatic man would have failed. So much so, in fact, that there was a complaint to the Board in 1887 that the Library suffered greatly by not having a Director to protect its interests. "The chief fault was the absence of a chief librarian in constant intercourse with the officials. Take the Museum, for instance, and all would confess that its director never lost an opportunity of getting a grant, no matter what suffered."[14]

This vexed question of funding led to considerable debate. Although the endowment for higher education was relatively straightforward, the second endowment led to problems of how to apportion the money. The battle between "blacksmiths and books"[15] raged for several years as various members attacked the Board for starving the library to the benefit of the Museum and the embryonic School of Technical Science, at that time merely the technological display in the Museum itself. The Board of Governors did not like criticism, even from its own members; it wriggled, it evaded. It claimed, for example, that the library mentioned in the original ordinance did not refer to the public library at all, but to quite a different library connected with the School of Technical Science. (The issue was always to do with the reference library, not the lending library which was supposed to be self-sufficient.) It even claimed thereby that the money which the Board devoted each year to the library was in the nature of an act of charity. Even a clear statement from William Rolleston, who had been the Superintendent of Canterbury in 1872 at the time of the ordinance, that it *was* the reference library that was referred to, "the centre from which scientific education and culture were to radiate,"[16] did not totally eliminate the Board's speciousness. It had to fall back on the justification that the money had been spent on the School of Technical Science, which not unnaturally produced the question "Well, where is it?"

The economic/political pressure on the Board of Governors and the constant publicity given to the issue of technical education by leading figures in Christchurch might not have been sufficient on their own to induce the Board to agree to such a venturesome undertaking if the growing engineering and industrial firms of the city had not put forward their claims. The Industrial Association, now known as the

Manufacturers Association, was formed in Christchurch on 12 August 1879 and concern was regularly expressed at its meetings at the lack of technical education for the many apprentices at the Addington Workshops, the Tramway Board and such notable firms as Andersons and Scott Bros. By 1884, concern had become so paramount that the Government was approached with the request that some kind of technical education should be made available in Christchurch as part of a national programme. The Government was sympathetic but fobbed off the Industrial Association with the suggestion that the Board of Governors was the best source administratively and, more to the point, financially for such a scheme. (The imaginative attempts of the provincial government to earmark the resources of Canterbury thus backfired, as indeed they were to do continuously throughout the years, as the central government was able to use the wealth of the Canterbury endowments as a justification for not contributing towards the educational projects of that province.) Thus recommended by the Government, a deputation from the Industrial Association, led by J.L.Scott,* of Scott Bros, applied to meet the Board of Governors; they conferred amiably on 25 November 1884. The Board, like the Government, was sympathetic, gave the deputation a good hearing, wished it could do more, mentioned proudly the approaching public lectures offered by Edward Dobson on Building and Construction, but regretted that lack of money prevented it from doing anything more energetic to assist the Association. A stalemate appeared to have been reached and, temporarily, even the criticism of the Board's treatment of the Public Library died down. (This issue was revived sporadically over the years and it must have been with great relief that eventually the Board was able to hand over the management of the library to the City Council in 1948.)

But the question of technical education was still very alive, in the constant publicity given by the *Lyttelton Times*, for example, to the Industrial Exhibition of 1884, the careful and lovingly detailed description of the various displays and the general atmosphere of intent interest and enthusiasm. "Machinery displayed in motion is a popular part of the present Industrial Exhibition . . . When the whirr of Messrs Scott's

*Scott, John Lee, 1844-1913; b Derby, England; apprenticed as pattern maker and engineer at Atlas Works, Derby; arr NZ 1870; founded Scott Bros, ChCh, 1874, builders of steam engines, locomotives, electric lighting plants, and, in later years, Atlas coal and electric ranges; m Sydenham Borough Council, pres Canterbury Industrial Association; m BG, 1895-7; 1901-3; 1904-13.

steam engine, or the clash of the Kaiapoi loom is heard in the avenue . . . [there are crowds of] eager, enquiring sightseers."[17] At a meeting of the New Zealand Institute (alias the Philosophic Institute) on 3 April 1884, C.W.Purnell speaking on the subject of " . . . a System of Technical Education for Artizans" announced that "the time has arrived when active steps should be taken towards opening technical schools in Christchurch and the other large towns of the colony."[18] Even more persuasively, the Prime Minister and Minister of Education, Sir Robert Stout, in his speech, "Public Education in New Zealand", on 18 June 1886 endeavoured to scotch the provincial rivalry that was so much a feature of these early days of New Zealand's history, by recommending that the various university colleges should specialise:

> When . . . the Midland Railway opens up the vast mineral deposits of the Middle Island, there might be established at Christchurch a School of Mines that would have a professor of geology and mineralogy (including petrology), a professor of surveying, and a professor of what has been termed natural philosophy, mechanics etc.

(This sounds clear-cut enough until we direct our attention to his suggestions for Auckland which quite unconsciously foreshadow the bitter rivalry between the Canterbury and Auckland Schools of Engineering, when for Auckland he recommended "an observatory and a professor of astronomy, and professors dealing with mechanical engineering, with navigation, with physical geography, with also perhaps, marine and land surveying, and civil and marine engineering. It would also have technical schools allied to it, considering the development that has been made in Auckland and in industrial enterprise."[19])

Meanwhile a great deal of pressure and politicking had inevitably gone on off-stage. After all, the population of Christchurch at this time was very small — 44,688 in the 1886 census — and the leading actors in this curtain-raising drama, judging from Heinrich von Haast's biography of his father, all knew one another socially. Much conferring undoubtedly took place amongst the prominent men of the city, the members of the Board of Governors, Julius von Haast, and his father-in-law, Edward Dobson, who not only was deservedly the grand old man of Canterbury engineering — with such notable achievements as the Lyttelton Tunnel to his credit — but was still employed by the Board from time to time as a consultant engineer and surveyor. Thus it seemed logical when the question of its possible responsibility towards providing engineering education was raised once more for the Board to request Edward Dobson to submit a proposal for its March meeting, 1887. Dobson complied immediately. His accompanying letter stated: "I forward herewith a sketch of a scheme for the establishment of a school of engineering which I think you will find comprehensive and which can be carried out gradually without much initial expense beyond the establishment of a chair of engineering."

If the Board had been in a position to realise Dobson's proposals, the School would have demonstrated a marked civil engineering/surveying/architecture bias in contradistinction to the mechanical engineering bias it was actually to acquire under Scott. Dobson proposed offering a diploma of the "associateship of the School of Engineering" stressing that "although the lectures are open to the public on payment of the fees, matriculated students only will be admitted to examinations or granted certificates." At least 12 months' practical work was specified "to the satisfaction of his employer in the special branch to which he has devoted his attention." As well he stipulated that a student should have "a knowledge of Latin and of one modern language besides English, either French or German, to the extent required for the B.A. degree." Civil engineering, architecture, and mechanical engineering were the three specialist course options for the second and third year of the diploma following a general introductory year of mathematics, elementary physics and mechanics, geometrical and freehand drawing, and elementary mensuration and surveying. Dobson provided an accompanying table which set out the distribution of teaching power that would meet the above requirements: Professors Cook, Bickerton, von Haast from Canterbury College and Messrs Scott and Elliot from the School of Art. In the immediate future he foresaw the need to appoint an Engineering Professor, and later, when students had reached their third year, he indicated that the services of a professional architect would be needed. Otherwise his requirements were modest: separate lecture rooms for the civil and mechanical engineering branches — he anticipated that architecture would be taught at the School of Art — and a small room to house the models. In the light of later complaints that graduates of Canterbury were pitifully ill-prepared because of the lack of workshop facilities, it is interesting to read Dobson's opinion that "an important adjunct to the School would be a workshop under the superintendence of one or more skilled mechanics, in which the student would be taught the use of tools,"[20] a use stigmatised by a Dean a generation later, as "handicrafts."[21] His initial premise, therefore, was that the School of Engineering would serve the needs of the professional class and would cater for matriculated students who had Latin and one other language. He was offering, in effect, the outline for a degree in engineering, although, as Canterbury College was legally empowered only to teach and not to examine,* this would require negotiations with the University of New Zealand and the passing of a parliamentary bill.

At this time, the chairman of the Board of Governors was F. de C. Malet, a man "who had a high reputation as a business man and administrator."[22] As he said himself, he took a strong "interest in the establishment" of engineering at Canterbury. Malet, while fully appreciating the value of what Dobson had proposed, was very aware that, at this stage, the Board could really only offer something far less pretentious.

> The proposal . . . to appoint a lecturer in the meantime is one which I assume will commend itself to the Board. It's more than likely the number of students at the outset will not warrant the cost of the appointment of a professor, but those who have been associated with the College from the first, and have seen how the various classes have increased in number, will fully understand that, as time goes on, and the numbers increase, the appointment of a professor will become both advisable and necessary.

Malet indicated that by appointing a lecturer in engineering and utilising the teaching already provided both in the College and at the School of Art, it was possible to "unite the two into a well-appointed School of Engineering, which would add an entirely new and useful department to the College for a comparatively nominal expenditure." His whole argument was pragmatic and low-key, apart from a skilful reference to the the emotive issue of provincial rivalry. "I may here incidentally remind the Board that the University of Otago some years back established a School of Mines and a School of Medicine at Dunedin. I am aware that the question of a School of Engineering has been mooted, and it is only a question of time — and probably only a short time — when we shall see steps taken with that object in another part of the

*see p. 33

Colony, unless we move in the matter." Canterbury had already lost the School of Mines; it was most unwilling for Otago once more to pre-empt it, special schools attracting prestige and students and money. At this point in the discussion, Malet moved "that it might be more convenient for the Board to go into Committee to consider the [matter]."[23] The Board in camera adopted the proposal that a School of Engineering should be established at Canterbury.

The news of the Board of Governors' resolution to establish a School of Engineering was greeted enthusiastically by the *Press*, and with great indignation by the *Lyttelton Times*, which deplored the closed meeting and felt that it was, anyway, too ambitious a scheme for so young a colony; certainly technical education was needed but other things were needed more. The *Times* declared:

> We regret the action of the Board of Governors on Monday; we regret both what they did and the manner of its doing. It appears to us that they not only did the wrong thing, but did it in the wrong way. The establishment of a School of Engineering is far too important a matter to be finally resolved on without full and open discussion. Such discussion did not take place on Monday. The Governors — ostrich-like — hid their heads in Committee. They must, therefore, not complain if outsiders do not think the attitude dignified. The ostrich, with its head in the sand, is not a dignified bird. The Board must have been aware that very great difference of opinion exists outside their own body on the advisability of establishing a department of engineering at the College. For ourselves, we believe that the weight of opinion is against so expensive and premature an institution. People think that too much money is being given to a small section in the community. Moreover there is the question of *cui bono*? What good will a school do which will merely increase a class far too numerous already?[24]

Outwardly impervious, the Board went ahead with its plans. It had already been decided that R.J.Scott, even though at this stage College minutes confused him with J.L.Scott, should continue to give the lectures on the Principles of the Steam Engine and Geometric Drawing which he had previously given under the auspices of the School of Art. (The fact that R.J.Scott was a man very sure of his rights and very ready to claim them was immediately apparent in a brief epistolary exchange between him and the Board over the question of his agreed salary, which led to a Board resolution that his salary of £150 should be backdated as from January 1887.) For a short time nothing was announced about a second lecturer. The Board received applications from George Phipps Williams, former District Engineer of the Canterbury branch of the Public Works Department, and from Edward Dobson, complete with curriculum vitae, and from our vantage point it seems a foregone conclusion that Dobson would be successful. It certainly appeared thus to George Phipps Williams, who wrote to the Board impugning their method of selection; he received a dusty letter for his pains, strongly repudiating such a notion and claiming that the same procedures as usual were observed. As the College was still in its infancy and there was a dearth of part-time lecturers, there had hardly been time to establish the kind of precedent the Board was claiming, but no more was heard on the subject from George Phipps Williams, although he was later to be closely associated with the School, offering lectures on Harbour Engineering. Edward Dobson was appointed at £150 per annum to provide lectures on Surveying and the Principles of Civil Engineering. The Board asked Cook, the mathematics professor, to supervise the running of the School, as neither of the two lecturers, Dobson or Scott, was employed full-time. Despite reservations from various members of the public, that "there is no demand for the creation of such a department of instruction, and delusive hopes will inevitably be fostered,"[25] the School of Engineering was launched with the inaugural address of 26 July 1887.

The annual report of the Board of Governors in 1888 listed the numbers of students attending the four offered classes:

Civil Engineering:	land surveying and levelling — 3
	principles of civil engineering — 3
	building construction — 3
	field instruction in surveying — 3
Mechanical Engineering:	freehand and mechanical drawing — 19
	practical geometry and mechanical drawing — 19
	lectures on the steam engine — 18[26]

It would seem that mechanical engineering far outstripped civil engineering in its appeal to the youth of Christchurch. This most likely reflects a vocational bias with the majority of the students coming from local industry; it must also, in part reflect the attraction of the steam engine in itself. Anyway, the numbers, for many years, remained consistent to this bias. At this stage, all that was offered were the evening classes for the apprentices of Christchurch. The complicated negotations that were necessary to persuade the Senate of the University of New Zealand to accept a degree in engineering were still in the future. For its first years, the School of Engineering could only offer evening classes to the local youth in "a small room vacated by the College Library"[27] and with no special equipment.

The two founding lecturers at the School, Edward Dobson and Robert Julian Scott presented an extraordinary contrast.

> Edward Dobson was a regular John Bull with his sterling character and reserve. His countenance might have been that of an English country squire; a high brow, a straight nose, a firm mouth, well-rounded cheeks, and a fresh complexion. He was clean-shaven except for reddish side-whiskers. In his eyes determination was mingled with kindness. He had the virtues of the Pilgrim Fathers, their sturdiness and uprightness, and something of their rigidity and lack of tolerance for the failings of their weaker brethren. He was a strong man who knew his own mind and spoke it. Once determined upon a course of action, he could not be moved from it by any thought of political expediency. He was not only a highly capable practical engineer, reliable — as Moorhouse knew when he said he would swear by Dobson's estimates — of vigorous initiative and driving power as his rapid construction of the road down the Otira Gorge in the depths of winter [revealed], but a ripe scholar, who had studied the classical works, architectural and engineering, of the ancients to some purpose, as he showed in the popular lectures he gave to the people of Canterbury.[28]

"The great old pioneer surveyor . . . a picturesque figure in Worcester Street, with his long coat and large bell-topper"[29] was at this time in his seventies and semi-retired; his great achievements in connection with the major public works of Canterbury had already become a matter of history. Robert Julian Scott, by contrast, had "the 'quarter-deck' manner."[30] At the time of his appointment as part-time lecturer, he was at the peak of his career and energy, working on designs for New

Zealand built locomotives. He was a man of "unique personality"[31], and from the time of his appointment to the position of lecturer in charge in 1890, until his retirement in 1923, the fate of the School of Engineering was inextricably rivetted to his life and character. Indeed, so great was his influence, that in the succeeding years until the move out to Ilam, the School was still dominated by the memory of the great man, enshrined in the courses and preserved in the teaching styles of staff members trained either by him, or by those who had taught with him.

He was born in Plymouth in 1861, the son of a Rear Admiral, and first cousin to Captain Robert Falcon Scott, the Antarctic explorer. The sea was in his blood and after his arrival in New Zealand he was an enthusiastic and expert sailor. There is still a Scott Trophy offered each year by the Banks Peninsula Cruising Club. Scott designed and raced ocean going yachts and great seamen's tales are told of his feats of daring and endurance. One special episode, which Scott himself described as "one of [his] wildest yachting experiences" concerned "an ocean race with a southerly buster blowing" when his yacht *Yvonne* "was carried over the sea-wall at the Otago heads."

> Suddenly out of the dust and driving spray the *Iarangi*, equally out of control, swept across our bows, missing a head on collision by a few feet. Then came a wild cry from those for'd "The Mole!" Though by this time the peak purchase had been cast off and the mainsheet eased it was evident that she was not paying off fast enough to clear the mole, over which a tremendous sea was now breaking. At the last moment, with I think, some idea of running the vessel up as far as possible I put the helm hard down, and struck heavily square on to the mole. Her head rose high in the sea that broke overfor'd. This was followed by another sea amidships but hove down by the wind on one quarter and thrown ahead by the break on the other she still drove on. "Glory be to God, she ain't stopped yet!" yelled old Sam. Then a sea broke over aft completely smothering us; her bows disappeared and her stern rose high in the air and she suddenly came up on an almost even keel. She had taken her seven feet of draft across 50 feet of rubble wall awash in the breaking sea![32]

"Loppy", as Scott was called because of his heavy ungainly walk, was immensely popular with his students, and he was "especially . . . loved by those who knew him best,"[33] the small band of the students of the '90s. In the early days of the School it was common for staff and students to be invited to cruise with him round Banks Peninsula. When he went overseas to recover his health in 1904, many students turned up at the Christchurch Railway Station to see him safely on board the boat train to Lyttelton. On leaving Wellington, he cabled farewell greetings to students and staff.

Scott's dash and daring at sea, his authoritarian manner, the great bulk of the man — he was very tall and broad and his eating and drinking were commensurate with his size — all combined to create numbers of Scott stories. Whether these are apocryphal or not, they still tell us something of his quality. One favourite which used to circulate the Lyttelton docks was of his self-designed motorised yacht breaking down somewhere out in the stream. For all his vaunted mechanical skills he was unable to discover what was the matter with his craft. He had, ignominiously, to be towed back into harbour. Or there was the time when he was sailing *Yvonne*, and met up with a fishing trawler in a raging southerly off the Lyttelton Heads. He challenged the trawler's West Indian captain to race him into harbour; and to his chagrin he was beaten. If the Board of Governors found it strange to address communications to Professor Scott, *Zephyr*, Port Chalmers, or receive replies stating that Scott would be delayed in his return to the College because of adverse winds, it did not show it.

Later in his life, he had a house on Horomaungu Island in Port Levy, which contained many great cushions scattered about so that he could rest his bulk at need. In 1889, he had married Gertrude, eldest daughter of C.C.Bowen.* They had no children and she died in 1909, at a time when Scott himself was ill in Dunedin, whither he always repaired when he needed expert medical attention. The fact that he was unable to return for the funeral inspired a very concerned letter of commiseration, hoping he was "bearing up bravely"[34], from the Chairman of the Board, the Hon. G.W.Russell.

Scott's enthusiasm for engineering design was conspicuous at an early

*Bowen, Sir Charles Christopher, 1830-1917; magistrate, politician and educational reformer. educ Rugby and Cambridge; secretary to J.R.Godley, 1850-2; Resident Magistrate in ChCh, 1864-74; president Canty Collegiate Union, 1872-4; MLC, 1874, 1891-1917; MHR, 1875-81; Min, 1874-7; m BG, 1873-5; m Sen UNZ, 1881-2, 1888-1915; Kt 1910; K.C.M.G., 1913.

age. In an address to the Canterbury College Dialectic Society in 1890, he described his early experiences with a four-wheeled velocipede which he had owned "in the early seventies. To propel it on the level required more exertion than to walk the same distance; it was best pulled up hill by a strong rope and the assistance of small boys of the neighbourhood, but it ran downhill finely, with great noise and *éclat.* As an improvement, in 1875, [he] designed and had built a tricycle with two equal-sized front wheels (the driving wheels) and a single rear steering wheel. This was the prototype of machines made many years later by various Coventry firms . . . "[35] He attended the London School of Mines, and then, before leaving England, served for a time with the Locomotive Department of the London, Brighton and South Coast Railway. On arrival in New Zealand, he joined the Railway Service where he rose to the rank of General Manager of the Addington Workshops, engaged primarily in locomotive design and manufacture. At the time of his appointment to the School of Engineering, he was already regarded as an engineer of distinction:

> In 1880 he designed and operated in Dunedin the first motor-car in New Zealand or Australia, a 35 h.p. steam buggy. In 1881 he designed for the New Zealand railways the first insulated waggon for the carriage of frozen meat, of the type that [was to be] universally used as the standard. He prepared the drawings and specifications of the first locomotive built in New Zealand (class D, built by Scott Bros., Christchurch), and designed and built at Addington the first locomotive made by the New Zealand Government Railways. He originated the "Prairie" type of locomotive, originally termed in America the "Scott" type. The first oil engines made in New Zealand by Andersons were designed by him.[36]

Scott was immensely ambitious; to have given up his promising career with the New Zealand Railways he must have indeed been given a firm surety of promotion in status and salary if he made the School succeed. The School rapidly became identical in his mind with himself. One very prominent characteristic of the man was his touchy self-protectiveness if he felt his dignity was being slighted. Any possible hint of criticism of the School was likewise not well received and Scott would burst into print, writing long and carefully detailed letters of rebuttal to the Board. He had no skill in delegating, and would not have been able to tolerate the much more democratic organisation of the School which exists today. His strong notions as to the credit deserving both to himself and the School, while they might make him a difficult colleague, certainly contributed greatly to the School's soaring success over these early years. A story told against Scott describes an occasion in the staff common room when one of his subordinates presumed to ask him about some detail concerning the School's organisation. He delivered a resounding snub: "I never discuss such matters with anyone below the rank of professor."[37] This "academic admiral"[38], the sole professor of the School of Engineering for 36 years, was undisputed master of his ship.

CHAPTER TWO

School of Engineering: "Humble Beginnings"

Although the Board's original decision in 1887 had been to establish a chair of engineering, a couple of years passed with no more significant action than the appointment of the two part-time lecturers. It is questionable how long such a state of affairs would have lasted if Scott, in 1889, had not been promoted to the position of Head Office Engineer at Wellington. Consternation reigned as there was no one available to teach the courses he had devised and which had appeared in the Canterbury College calendar for 1890. To withdraw Scott's classes and leave only the far less popular classes given by Edward Dobson would make a mockery of the newly established Canterbury School of Engineering. Scott had already required six weeks' leave of absence in 1889 because of his Railways work. As a result, the second term of that year had been in danger of being suspended until J.L.Scott of Scott Bros., offered to stand in free of charge. The Board thanked him gratefully for his offer, but, nonetheless, paid him £21. (The willingness showed both by him and Arthur Dudley Dobson* to assist the School in any way practicable is evidence of the close links between practising engineers and the School which were forged way back in its pioneering days and remain alive today.) With Professor Cook the nominal head of the School, it was just possible to accommodate a temporary absence. It would be a different matter if Scott were to resign altogether. The Board was only too well aware that he was the only suitable man to teach the classes in Mechanical Engineering. When it had consulted him initially about the possibility of his teaching at the School of Art,

*Dobson, Arthur Dudley, 1841-1939; second son of Edward Dobson; arr Canterbury 1850; educ Tasmania and Christ's College; trained as civil engineer under Edward Dobson; ChCh City Engineer 1901-21; pres NZ Soc of Civil Engineers 1925; knighted for services to engineering 1931. A pioneer par excellence, he assisted von Haast in geological surveys, drained the Rangiora swamp, discovered Arthurs Pass and like his father, held posts both as consultant and official.

he had declined. He was requested to suggest a possible lecturer and it was because he was unable to recommend anyone else that he had reluctantly agreed to take the classes himself.

On learning of his promotion to Wellington, Scott immediately communicated the situation to both Cook and Malet. The Board's response was to request Malet to visit Australia in order that he might perceive at first hand how the two engineering schools of Melbourne and Sydney dealt with their problems of finance, buildings, apparatus, staffing and so on. It was on his recommendation that the Board offered Scott the position of lecturer in charge, "the lectureship . . . to be developed into a Chair when the success of the School warranted it."[1] A salary of £550 was agreed upon because it was the amount Scott was currently commanding at Dunedin, whither he had been transferred pending his acceptance of the appointment at Wellington. From his own later account of this agreement, he did not press for the higher salary commensurate with his qualifications and status, because to do so might well have prejudiced the establishment of the promised engineering laboratory. As a result of various discussions, he "declined the proffered promotion," and expressed his "willingness to remain in the Board's service for the purpose of founding an Engineering Laboratory, and taking charge of the Engineering Department."[2] Scott was firmly established in the Railways. Yet he chose to abandon this impressive and promising career and commit himself to the School of Engineering. Thus started his long fight to give the School not just provincial, but national, status and recognition.

In his opening address for the year 1890 Scott offered a peremptory account of his vision for the School:

> Future engineers should receive a good general education, should then acquire a knowledge of mathematics, chemistry, physics, mechanics, the nature and strength of materials, the theory of the steam engine, and drawing. They should then have two years' practical education in the workshops, and at least one year in office or field work. In connection with the Board of Governors' School of Technical Training, there would be a laboratory, provided with appliances for testing the strength of materials, an experimental engine and boiler, besides drawing and lecture rooms. Tests of the value of fuels could be carried out, and the calorific and evaporating power of our coals accurately determined. A plant for the investigation of physical phenomena and

> electric lighting would also find a place in the laboratory . . . The whole community were beginning to recognise how much the safety of life and limb was dependent on the skill and intelligence of our mechanics. The efficiency of machinery, and of many important public works, was largely dependent upon them. There was also ample scope for New Zealand engineers trained to be familiar with the resources and requirements of our country. Australia had already shown that she could train her own engineers successfully. Why not New Zealand?[3]

"Scott's genius is best shown by the early recognition of the academic trend of engineering education. With some vision, he saw the intrusion of the training of engineers into the University sphere."[4] But, at the time of his appointment to the position of lecturer in charge, there was no engineering degree, and it was not within the powers of the Board of Governors to establish any such. For Canterbury College, and the other provincial University colleges of New Zealand, were empowered merely to teach, not to examine. According to an Act of Parliament passed on 31 August 1874, it was the prerogative of the University of New Zealand to "appoint examiners, conduct annual examinations, and to confer degrees."[5] If Canterbury's prestigious special school was to offer degrees in engineering, approval had to be granted by the Senate of the University of New Zealand. It was fortunate that Sir Robert Stout wholeheartedly supported Canterbury's proposal. He was greatly in favour of the different affiliated colleges of the University of New Zealand diversifying into distinct specialisms. From the very beginning of the School's history, he pressed the Senate to acknowledge the graduate courses.

There was initially a certain reluctance; the Senate must have felt it was creating a precedent. After all there was no centuries-old tradition of engineering schools in the Universities of the United Kingdom; they were a relatively recent innovation. And the men who sat on the Senate were not in a position to regard themselves as authorities on engineering. They could, and did, argue at some length about the suitability of certain Latin texts as set books for classes which contained women as well as men. Everyone could become an instant expert in such areas. But deciding between the different components of a degree in engineering was another matter.

Stout's first move was rejected. He persevered; the Senate grudgingly agreed to offer engineering certificates. In the annual report on behalf of the School of Engineering which Scott submitted to the Board in 1893, he made it very clear that he found this eminently unsatisfactory. "A diminution in the number of matriculated students has been caused by the impracticable nature of the course of engineering at first decided on by the University Senate, the reward for the completion of which was to have been a certificate only." He also felt that the fact that there was the same number of matriculated students attending the School two years running — 8 as against 45 non-matriculated students — called for some justification, in case anyone held it against the School and by implication against him. Fortunately, the certificate was a temporary measure. The Senate's unfavourable vote of 27 February 1892 was reversed a few days later, and henceforth a B.Sc.(Eng) was offered in all three branches of engineering. Thus Scott could proceed more cheerfully: "the Senate . . . has been prevailed on to reconsider its decision; and the course, as now laid down, is an excellent one."[6] The tone of self-congratulation was very marked; if the course laid down was an excellent one, then full credit must be given to Scott himself. In 1890, the Professorial Council recommending the proposed engineering syllabus to the Board of Governors submitted Scott's 27 handwritten pages. The various aspects of the mechanical engineering course were presented in great detail. In comparison, the few pages in Dobson's handwriting devoted to the civil engineering component of the offered courses were remarkably restrained. In *Early New Zealand Engineers*, W.F.Furkert, Commissioner of Works, and, for several years after Scott's retirement, a member of the Engineering Faculty, gives full credit to Scott for his achievement in establishing the engineering degree, declaring that "the syllabus and course for the Engineering Degree of Canterbury was entirely Scott's work, and in its comprehensiveness, broadness and insistence on essentials [was] equalled probably only by that of Cambridge in the English-speaking world."[7]

Scott's eagerness to establish the engineering laboratory was initially checked by the Board's reluctance to commit itself to any great expenditure. Malet's recommendation on his return from Australia had been persuasive. "I venture to think that we may safely follow the example of [Victoria and New South Wales] and proceed cautiously, developing the department as the success of the school warrants and our means admit." He certainly recognised that there would have to be some capital expenditure to provide plant and then an appropriate building to contain it but was hopeful that "a total outlay of about £1500, extended over a period of say four years, would . . . be found sufficient for the efficient working of the department for the years to come . . . To provide for the proper accommodation of the students, whose numbers are increasing, the Board has determined to erect additions to the College, with such facilities for the special needs of the school as the experience of the lecturer in charge might suggest."[8] This work, he reported, was already in hand. The Mechanical Laboratory, designed by B.W.Mountfort, was completed in 1891 at the cost of £2,740. But while the necessary building was erected promptly, and Scott in his role of lecturer in charge was consulted as to its most advantageous lay-out, he was still faced with considerable delays before the apparatus was authorised. The Board approved of the engineering laboratory in principle but was very reluctant to authorise the necessary expenditure.

On 1 February 1891, Scott optimistically submitted estimates for the plant he required for the School, with a recommendation that Professor Alex B.W.Kennedy (later Sir Alex) "be requested to generally supervise [its] construction." He had done "similar work at extremely moderate fees for the Australian Universities and with most satisfactory results."[9] The plant which Scott wished the Board to order consisted of a testing machine (£685), a lathe (£75) and an 8 H.P. experimental engine and boiler (£500). The end of that year, however, he wrote to the Board complaining at the lack of the necessary apparatus without which "it will be impossible to train the more advanced students in a manner satisfactory to myself or calculated to make them capable engineers." Scott warned that "the high position which the school has earned for itself in the estimate of members of the profession will be imperilled by the continued absence of all practical experimenting."[10] (Scott's recommendation that Professor Kennedy should be consulted indicates the extent to which he himself was *au fait* with the latest developments in engineering education. Kennedy was appointed to the Chair of Engineering, University College, London, in 1875; four years later he had established an engineering laboratory there, the first of its kind. This laboratory was described by him in a paper which he read before the British Institution of Civil Engineers in December 1886 and which was subsequently printed in the *Proceedings*. Kennedy stressed the importance of providing the student with the "opportunity of studying such things as the physical properties of the iron and steel . . . the strength of those materials . . . the efficiency of the machines . . . the relative economy of the different types of engines . . . the evaporative power of boilers." Such things are "to a wonderful extent, suited to set the student free from the thraldom of the engineering pocket-book (making 'every man his own Molesworth' . . .) by helping him to determine for himself, or at least to see practically how other people have determined, all the principal engineering constants, from the tenacity of wrought-iron to the calorific value of coal, or the discharge-coefficient of an orifice."[11] Scott would have greatly appreciated such a clear statement of the equipment essential for a well-appointed engineering laboratory.)

In his 1892 annual report Scott listed all the equipment at that time in possession of the School. It makes pitiful reading; equipment is too grandiose a term to describe what in effect was a set of tools:

> One 4½in centre screw cutting lathe, 6ft gap bed with overhead motion, chucks and wheelcutting appliances complete; one 4½in centre 3ft bed American wood-turning lathe complete; one 1¼in spindle drilling machine, which has been fitted with compound table and self-centring drill chuck; one Kennedy's patent shears, 1¼in blade; one hand shaper, 6in stroke with automatic feed and universal joints; one 6in instantaneous grip vice, with taper attachment; one gas hearth with tools, bellows and blow pipe complete; one anvil; one swage block; one set smith's tools; one high-power gas burner, with ladle etc; one complete set of hand tools (wood and iron) necessary for the preparation of test specimens, models etc . . .

Even so, the note of pride was unmistakable: "The laboratory has also been fitted with flooring, gas, benches, cupboards, etc. The tools obtained are all of the best quality. The presence of even this small amount of machinery has been of the greatest practical value in the instruction of the matriculated students. I trust," he continued tentatively, "that the finances of the Board will admit of its making additions from time to time of other appliances which are urgently needed, especially the testing and experimental appliances."[12] Under these circumstances, Scott lost no time in requisitioning the models from the Technical Department of the Museum, although this led to complaints. It had been possible under the old system for the public to get permission to examine these models closely and even carefully dismantle them, but once they became the property of the School of Engineering, this was no longer the case.

Still the Board made no move to authorise the purchase of the necessary plant. Scott's patience must have been sorely tried. But he was fortunate in that there were other pressures influencing the Board's decisions. For, in the meantime, Sir Robert Stout had battled successfully to persuade the Senate of the University of New Zealand that engineering subjects could be accredited to a B.Sc. degree. Included in the University statutes for this degree on Scott's recommendation and presented to the Senate by the Board's representative, Professor Cook, was a clause which demanded that a student follow a particular course of practical work, work which was actually not possible without the promised engineering plant. While Scott was immensely proud of "the great advantages of a University Course, and the official recognition of the School,"[13] he was becoming increasingly anxious at the Board's delay in equipping the School. In his 1893 annual report of the School of Engineering and Technical Science, he was once again vigorously complaining about the absence of "practical illustration", the result of New Zealand's isolation and the smallness of its population. If students were not to be seriously disadvantaged, the School of Engineering must offer "a well equipped laboratory," which inevitably necessitated "a considerable expenditure." His trump card, however, was the fact that "the University Senate . . . insist on a large amount of experimental work, and their examiners are to be 'English engineering professors of standing,' accustomed to this method of instruction."[14]

The Board, however, continued intransigent. Then in 1894, George Julius, (later Sir George), was prevented from sitting a final examination because he "had not qualified in experimental work, the necessary apparatus not being in existence in the College."[15] The publicity surrounding the Senate's refusal, no doubt assisted by the fact that Julius' father, Bishop Julius, was a prominent citizen and Board member gave Scott a very strong bargaining position, and he was asked to submit an itemised list of essential equipment. In addition to the equipment he had already requested he put in a claim for electrical engineering equipment to the value of £600, rightly anticipating that there would be a need before long to include electrical engineering in the curriculum. The Board was forced to concede the £1260, but was adamant that it could not provide the additional £600. The *Press* might fulminate against the "insatiable demands" of the School, "the hasty concession to a popular craze for technical education,"[16] a correspondent to the *Press* might suggest that the Board was "indulging in a fad"[17], but in fact the Board could not in all honour refuse to authorise the equipment when students following the offered courses in good faith were to be stopped at the finishing line because of lack of equipment. It is impossible not to sympathise with Scott's conciliatory and rather anxious request to Messrs Buckton and Co., Leeds, who were the suppliers of the testing machine that "as it is only with considerable difficulty that I have been able to obtain the necessary funds for these appliances I would ask you to make the most advantageous arrangements possible for freight etc."[18] The steam engine arrived in the middle of 1894, the testing machine not until 1895. There was no escaping the tyranny of distance. Then came the problem of customs dues, and with bureaucratic solemnity, the Customs Department wrote to the Board that there was "no provision for the admission of steam engines for educational purposes."[19] The Board wrote carefully explaining; Scott wrote equally carefully, but the Customs Department was adamant. Scott went to Wellington in the hopes of wheedling concessions out of the premier, Richard Seddon. His first telegram of 4 July 1894 read: "Believe shall settle matters satisfactorily having hard fight."[20] By 4.30 the same day he had won and the gleeful telegram ran: "Interviewed premier duty engine remitted testing machine under consideration."[21] In this way a precedent was established and the Customs Department adjusted itself to waiving customs dues on a variety of educational machines for the School of Engineering. Scott's tenacity and exuberance had saved the Board a considerable amount in customs dues but for all that it must have viewed gloomily its prospects of curbing the enthusiasm of the lecturer in charge. Scott was advised, 28 September 1894, "to keep within the estimates". But, despite the need for "stringent economies", there were more expenses for the Board as adequate foundations for the massive testing machine had to be provided; it seems allowance had not been made for the "well-known stratum of quicksand below Christchurch."[22] Scott Bros won the contract for erecting the machine, under Scott's inevitable supervision. H.R. Webb, the Chairman of the Board, sent a telegram authorising the erection, with the additional comment: "Tell Scott he must keep charges down."[23]

If the *Press* was a stern critic of this "impulsive hasty expenditure"[24], the *Times* expressed its approval by a full column description, on 10 September 1894, of the new steam engine:

> The experimental steam plant for the engineering department of Canterbury College, is an excellent one, and admirably suited for

its purpose, or rather for its purposes, for there are many of them, as will be seen from the following list of objects with which the plant was ordered:-

Illustrating the modern theory of heat; tracing the action of heat in the modern steam engine; enabling the best relative proportions of an engine for working under any given conditions to be experimentally ascertained by means of practical illustration of the effect of any possible alteration of such proportions; providing an apparatus for the exact testing of the calorific and commercial values of various fuels; illustrating lectures on the steam engine; instructing students in valve setting, designing and practical experiments.[25]

The *Times* was equally enthusiastic the following year about the testing machine, which conferred such advantages on the community:

> By means of these ["latest scientific appliances"] a twofold end is secured: first, instruction is imparted to students of practical engineering; second, which will perhaps appeal to many who may not perceive all the merits of a technical school, accurate determination is obtainable as to the physical nature of any crude or manufactured substance, such as metal, timber, fibrous materials, concrete, leather, rubber, etc. That is to say, anyone is now in a position to obtain accurate information as to the quality of his materials. The necessity of this has long been recognised in other countries, where, instead of a rule-of-thumb calculation by some irresponsible person, the dictum of a competent and responsible official as to the strength of material is compulsory.[26]

At a conversazione given under the auspices of the Science Society on 9 September 1895, "the engineering department was one of the most popular resorts of the evening, for the fine new engines and boiler, just erected, were working, and doing so very satisfactorily."[27] Once the testing machine had arrived and been installed, Scott submitted a schedule of testing fees to the Board. As it was his full responsibility to ensure that the tests were performed adequately, so that no imputation of inefficiency could damage the Board, he claimed that he was entitled to 50 per cent of the fees, the remainder going to swell the finances of the Engineering School. The Board, perforce, agreed, but not before Scott had felt there was some slur against his name, some suggestion that he was undertaking consultancy work without the Board's consent. The fluent manner in which he wrote to the Board in his own defence was to become very familiar over the years. He declared: "I have proposed regulations which would permit me to carry out work of great public utility under that clause of the Board's offer (November 20-89) which provides for my undertaking outside work with their consent, and under which I have by arrangement carried on a selected amount of consulting practice." He continued forcibly: "I would be greatly obliged by your correcting the erroneous impression which apparently exists at an early date."[28] The whole matter was ironed out, and various suggestions that the testing machine was not really of much practical value to the students were ignored.

By now Scott was into his stride. He had his engineering laboratory complete with equipment, his degree students as well as the evening students who continued to attend the School in great numbers. For, after his Australian visit, in addition to advising the Board to defer its decision to create a chair of engineering, Malet had strongly emphasised the need to establish technical education within the range of local apprentices, not just financially but academically. This had earned the approval of the *Times*, previously a vociferous critic of the potential expense of a School. Such evening classes would prove "cheap, practical, and likely to be generally useful"[29]. So, instead of a possible slow phasing out of the evening classes in favour of an associateship diploma or degree in engineering, there was a decisive acknowledgement that the School of Engineering should have a two-fold role. Little by little, it was to provide for the education of professional engineers, but in the early years the mainstay of its existence was to be the evening or "extra" students. It had been the needs of the non-matriculated students which had been espoused by the Industrial Association, and it was largely to them that the popular lectures of the previous decades had been addressed. The main objection offered to the establishment of a degree course was that too few students would benefit and that a colony the size of New Zealand, while it needed trained mechanics, could not absorb numbers of highly qualified men. Certainly the history of the railways in New Zealand shows how difficult it was to find suitably trained men within the colony to oversee the track once it was laid. Scott, accustomed as he was to working with apprentices at the Addington Workshops, accepted the two-fold function of the School. It is likely that without such concurrence the School might have suffered the same fate as the Auckland School of Mines, which had rapidly become defunct without public support and Government assistance. Later, the extra work and responsibility entailed by running the School under these two separate banners weighed down the staff like an incubus, leaving neither time nor energy for independent research, creating overcrowding, and probably adding to the difficulties of recruiting suitable staff. Teaching at such a low academic standard was inevitably a hardship for any man really involved in developing engineering education. But at this time, the School needed the technical school to survive; it was not until just before the second world war that the numbers of matriculated students equalled those of the evening students and the School assumed its primary function of training professional engineers. It says much for Scott's commitment to the cause of engineering education that he saw it as his mission to build up, strengthen and maintain both aspects of the School under his charge. There is no evidence that Scott ever resented the two-fold function of the School. His enthusiasm was for engineering education and he was not an élitist. He set about encouraging maximum attendance at the School, sending prospectuses to be displayed prominently at the Railway Workshops at Addington and Wellington. Although he wanted to make attendance at evening classes compulsory, he was not able to gain acceptance for such a scheme. When, in 1908, he discovered that the reason apprentices were defaulting was because the fees were too high, he negotiated with the Board for reduced fees for students who were committed to two or more classes, encouraging, at the same time, the local firms to pay the students' fees. His eagle eye was quick to see possible encroachments on to his territory; where courses at the School of Art duplicated what he provided he wrote a stern rebuke to the Board. Similarly, after the Christchurch Technical School had been established in 1907 he was most prompt to complain when it offered courses which doubled with those of the School. Looking back at the origins of the School from today's vantage point, when it has an international reputation and attracts considerable amounts of research money in addition to overseas scholars, it is well to be reminded that, in its very humble beginnings, it catered for the needs of apprentice engineers and artisans rather than for members of the engineering profession. In those days the presence on campus of so many non-matriculated and part-time students requiring special arrangements of teaching and timetabling did not seem anomalous. In the early years of Canterbury

College, such students were necessary to swell the numbers. On the whole, they attended lectures in order to gain matriculation so that they could proceed to take out their degrees. The difference with the non-matriculated students of the Engineering School was that this path was not open to them; they attended as extra students, and as extra students they stayed. In this way the two-fold nature of the School was firmly established and it was not until 1946 that it was possible to begin transferring the first year apprentices to the Christchurch Technical School. At this early stage in the School's development, the evening students were absolutely necessary both politically and economically. They qualified for a capitation grant from the Government. They were the Board's justification for spending the endowment money from the reserves set aside for the Museum, Public Library and a School of Technical Science, a substantial proof that the Board had acted responsibly towards the citizens of Canterbury. They earned the goodwill and co-operation of the growing engineering and industrial firms of Christchurch.

But this is to anticipate. In the meantime, Scott's 1895 report exuded his sense of achievement. The measure of Scott's success were the positions achieved by former students of the School:

> I have the honour to report on the work of the past year of the department under my charge. The number of students attending has increased from fifty-one to sixty-two and the number of hour attendances per week from 254 to 308. Six matriculated students sat for the University examination in Engineering. Five passed in their respective years, the sixth passing in the technical but failing in the collateral subjects. Fourteen students obtained the College certificate in freehand mechanical drawing, eight in mechanical drawing, five in descriptive geometry and setting out work, six in steam and one in applied mechanics. Since the foundation of the department 192 names have been placed on its books, 130 persons having completed courses of varying lengths. Of these, starting in nearly every case from the level of apprentices, two have attained to the position of engineers in charge of undertakings of magnitude, two that of works manager, two are lecturers and one a demonstrator in the service of the Board, nine are draughtsmen, four surveyors and draughtsmen, five shop foremen, eight engineers of sea going vessels, five engineers

of freezing works, pumping stations, etc., eight chargemen or leading mechanics, eight are in business on their own account. Fifty-four thus occupy positions of responsibility in England, Australia and the colony, fifty are working at their trades as skilled mechanics, the remaining twenty-six being made up of those whose present whereabouts are unknown or who have turned to other occupations. Such results form a complete answer to the question which has been raised as to the future employment of engineering students, and account for the steady increase of attendance which has been characteristic of the history of the technical school. I must, however, remind the Board that the department cannot be considered as fully equipped for modern requirements until provision has been made for a section of electrical engineering, the arguments in favour of which, advanced by me in 1890, apply with still greater force today.[30]

CHAPTER THREE

The First Decade

When the School was first established in 1887, the arrangement was that Professor Cook should undertake to supervise it, while Dobson and Scott shared equally between them the responsibilities of lecturing. It is obvious from the fact that Dobson gave the inaugural address, as well as the two addresses for the opening sessions in the years 1888 and 1889 respectively, that at this time he was regarded as the most senior engineer. Once Scott was appointed lecturer in charge this situation immediately changed. Dobson was totally overshadowed by Scott's meteoric career. He retained his position on the staff for a further three years, although there were very few students pursuing a course in civil engineering. Then, at the beginning of 1892, the minutes of the Museum and Library Committee warned that it might be necessary to make arrangements for a lecturer in surveying. If the Board, at Scott's recommendation, were intending to replace Dobson with a younger and more energetic man, this did not eventuate; nevertheless, at the end of that year Dobson was informed that "his services were no longer required."[1] The exquisite courtesy of the man is clearly revealed in the letter Dobson wrote to the Chairman of the Board on 26 December 1892, expressing his "gratification at the pleasant nature of the relations which have always existed with yourself and the other members of the Board during the time that I have been associated with the College." He concluded: "I shall always be pleased to place at the service of the Board of Governors any information or professional experience of mine, which may be of value to the work of the Technical School, in whose successful progress I shall always feel a great interest."[2] Edward Dobson's association with the School of Engineering had been a very short one, but, for all that, immensely significant. He was an old man with his great achievements behind him — the Lyttelton Tunnel, the

Main South Road and the Christchurch to Greymouth road. The School can be proud that he was so actively involved in its founding, when he can be seen as the grand old man of pioneer engineering.

With Dobson no longer teaching in the School, there was no provision for courses in civil engineering. It was not until 1904 that this branch of engineering was once more offered at the School. Dobson's departure freed £150 a year and Scott was not slow to earmark that amount. (The whole episode of Dobson's "resignation",[3] as Scott was later to refer to it, is most ambiguous. It is tempting to speculate that, in fact, Scott took advantage of the poor attendance at the civil engineering classes to dispense with so prestigious an engineer. No matter how circumspectly Dobson behaved, Scott would inevitably see him as a rival.) Obviously he could not carry the whole weight of the School unaided — the administrative work involved because of the two-fold function of the School was in itself an immense burden, without the teaching and supervision of drawing classes — and he recommended to the Board in 1893 that the position of demonstrator should be advertised. He submitted an impressive schedule of qualifications. The applicant should have spent a "period of not less than 5 years in Workshops and Drawing office," be "an expert draughtsman, a capable mechanic," with a "knowledge of elementary mathematics and mechanics at least equal to Junior Scholarship, and of descriptive geometry."[4] In the event, only one man applied. A.F.Morrison was an apprentice fitter, who had attended the evening classes at the School in 1890 and 1891. Potentially an excellent student, he obtained a Canterbury College Engineering Exhibition, and passed the College examinations. (The engineering degree was not recognised by the Senate until 1893.) In 1892, Morrison was transferred to the drawing office of the Chief Mechanical Engineer, New Zealand Railways, Wellington. He accepted the appointment to the School of Engineering, at a starting salary of £150 a year. Demonstrator was a rather ambiguous title for a job which included writing Scott's letters and being a general dogsbody. (Morrison was a better speller than his professor, who consistently misspelt such words as accommodation and vacation. His handwriting was also much neater and more precise; Scott's was large and flowing. An amusing characteristic of Scott's letters was that he never took the trouble to hyphenate a word; if he ran out of space at the end of line, he merely completed the word by running it down the right hand side of the page.) Scott and Morrison had a rather uneasy relationship. Morrison chafed at the menial nature of his position, but for the meantime he had to settle into the School and learn exactly what it was Scott required of him. His attempt to secure an appointment with the Wellington Harbour Board in 1898 was unsuccessful, as indeed, were other efforts he made to gain more satisfactory employment. And because of the top-heavy structure of University appointments, the position of demonstrator was a cul-de-sac. There was no built-in salary or promotion scale. Even if Morrison had been more qualified, Scott was firmly ensconced as god-professor; as a result, Morrison, somewhat fretfully, it is true, had to endure.

Meanwhile Scott himself was endeavouring to wring the promised promotion to the status of professor out of a recalcitrant Board. At the end of 1891, he presented the Board with an eloquent claim for it to review, as promised, his salary and status, but he must have realised that his timing was not appropriate because he wrote, early the following year, asking, that his letter should be withdrawn from the records. No explanation was offered for this withdrawal. That there was some degree of hostility towards his appointment becomes apparent from a charge of nepotism in the *Press*, 4 June 1892, when a correspondent wrote: "It makes one's blood boil to think that . . . the professor of Engineering, the son-in-law of another 'bosom friend' of the Chairman should get £550 per annum for lecturing to about three students."[5] It was possible to offer a smug repudiation of this accusation by inviting the correspondent to visit the School any weekday evening and see for himself the numbers of students attending lectures, and the serious charge of nepotism could be safely ignored. It seems likely, however, that there would have been an outcry if Scott had pursued his claim for a higher salary at this time.

Although not a professor, Scott had been granted a seat on the Professorial Council on 27 May 1890, and served as chairman in 1893. The Council at a meeting of 28 August 1894, which Scott himself did not attend, recommended that all lecturers in charge of departments should henceforth be designated professor. It was in this way that Scott became professor in charge; if at first it were an honorary title, over the years this fact came to be forgotten. There is absolutely no doubt that Scott was deserving of the status he so greatly desired but it is ironic that until he was created Professor Emeritus after his retirement, he could not, in strict accuracy, claim possession of a chair. He was certainly very sensitive about his title which suggests that there was a considerable amount of academic snobbery about. He had no degree and engineering was a very recent arrival on the University scene. Apparently he set such store by his title that some of his students decided to take the Micky out of him. They approached him with a request that he sit for a portrait by a student of the School of Art. He graciously agreed; the portrait was completed and displayed. A veritable professor in charge, tie askew, none too steady and firmly "in charge" between two constables. The offending portrait was removed although an empty frame survived to tell its own story. The artist was David Low.*

If Scott had decided in 1892 that it was not an appropriate time to push the Board to reconsider his status and salary, by 1897, after he had been associated with the School for 10 years, he was far less tentative. 1896 had seen the first graduates, B.Sc. (Eng) of the School — G.Julius and J.Parr — and accordingly he wrote to the Board asking, most persuasively, that it grant him the raise in both salary and status which he so desired. The Board, by way of A.C.Wilson, the Registrar, wrote repressively: "not under the circumstances." The letter began with the customary "Sir", and ended with the full formality of "I have the honour to be, Sir, your most obedient servant"[6] in marked contrast to the easy informality of an earlier letter from Wilson to Scott on a less ticklish subject which opened with "My dear Scott" and concluded with "Yours truly."[7] It is hardly surprising that Scott found the answer — "not under the circumstances" — far from satisfactory. His reply was taut and incisive. "May I ask what the circumstances alluded to are," he enquired, "for I am not aware of any circumstances which can be held to cancel the Board's obligations to myself. From the action of the Board, it has occurred to me, that the majority of its members may not be as fully acquainted with the nature of these obligations as I believed when writing my previous letter." Scott therefore reminded the Board of the interview he had had with Malet at which, as he explained, "I was asked by the Chairman, to state the salary at which I would accept this Lectureship, and he recommended me, in view of the great expense to be faced in the establishment of the Engineering Laboratory, to name as low an amount as possible, pointing out that with the expansion of the department there should be no difficulty in obtaining an increase." As a final, most telling point he concluded that:

*Although it is not possible to give a precise date for the portrait, it was executed during the middle period of Scott's leadership of the School. David Low began attending evening lectures at the School of Art in 1905.

"I would remind the Board that I was specially invited by them, to undertake the work of building up an Engineering Department at the College, presumably as they considered me the best qualified person to do so, and as an inducement I was led to expect a reward in the shape of position and salary, without the prospect of which my services would not have been available. I have fulfilled my part of this mutual understanding, and I now venture to ask the Board to fulfill theirs."[8]

Although the Board maintained an outwardly dignified silence, it did feel it should consult Professor Cook, who had been in on the original decision-making. After reading Scott's two letters, Cook replied to the Board: "My distinct recollection of the impression produced on my mind at that interview, is that it was understood that, if the school succeeded the lectureship should be raised to a professorship with the corresponding increase in income." Always a supporter of Scott, he referred to him as "a man who brought to [the School] enthusiasm as well as knowledge."[9] He had already expressed his belief that "the School of Engineering is destined to be a most important institution in the Colony,"[10] when he wrote to the Board thanking it for the memento he had been given in 1889, when he was relieved of the responsibility of supervising the School. For the moment, Scott did not pursue his grievance against the Board, and certainly it treated him most generously in 1899 when he was granted a year's leave of absence on full pay in order that he could visit an eminent eye specialist in Germany.

Even with the additional assistance provided by Morrison, the load on Scott was severe and he was forced to request that the School postpone the opening of its second session in 1896 for two weeks because he was in Dunedin seeking treatment for eye strain. (At that time in the College's history, the academic year was divided into only two terms, the three term system not coming into effect until 1921.) By 28 July of the same year Scott was, "with regret" asking the Board for additional assistance. A combination of the "hard, glaring light" and "the greatly increased number of evening students" meant that supervising evening drawing classes had brought on a severe inflammation of the eyes. The letter requesting such assistance was written from Wain's Hotel, Dunedin; Scott maintained he was "not allowed to read or write, [was] wearing coloured glasses, and [had] one eye completely closed." His suggestion as to how the work of the School should be carried out for the rest of the term provided for J.Parr, one of the first two graduates, to supervise the drawing classes on Monday and Wednesday nights. If his progress were to be delayed too long, Scott felt that Parr could undertake to read his lectures for him, as they are "written in full and fully notated." The letter was inscribed "I agree"[11] and signed H.R.Webb, the current Chairman of the Board.

Scott's eye strain cleared up temporarily but in 1897, the Board appointed as a part-time assistant an Austrian engineer, Otto Peez, whose application was accompanied by very impressive testimonials. The whole episode of Peez's appointment has its lighter side as Scott, on one of his yachting excursions down the south coast, was unable to get back to the College to take part in the selection. His explanatory telegram read: "Weather bound Port Chalmers"[12] and continued with a request that the Board send details of the applicants down to him. Later Scott sent a letter from Timaru recommending Peez's appointment because, as he explained: "I may not be able to get to Christchurch (having been delayed by contrary winds)."[13] By this time, the Board was obviously fully accustomed to the behaviour of its yachting professor and appointed Otto Peez without demur. On Scott's advice, Welsbach Incandescent lights were installed for the 1897 session.

But Scott's eyes continued to give trouble and by the beginning of 1899, he was back in Dunedin at Miss Cronach's Private Hospital where his medical advisor, Dr Ferguson had diagnosed "rheumatic iritis" and recommended a spell off work. Applying for a year's leave of absence, Scott, always an opportunist, undertook that as soon as his eyes were rested, he would devote his time overseas to promoting the interests of the School of Engineering. "I propose," he declared, "to utilise this leave in visiting Great Britain, and intend, as far as my very limited resources will permit, making myself familiar with the modern advances in engineering and methods adopted in engineering colleges."[14] The arrangements suggested by Scott for the running of the department in his absence were adopted by the Board, Cook returning once more to the role of supervisor, Arthur Dudley Dobson, at a salary of £200, becoming *locum tenens* and reading Scott's lectures. A certain anxiety was expressed about these lectures which were to be handed over to Dobson "of course without prejudice to [Scott's] copyright in them."[15] Once they had been delivered they were to be returned and stored safely; Dobson, who was "a delightful man, gentle and courteous,"[16] agreed and also accepted the stricture that he must assert adequate control of his classes. "I must impress upon you," he was informed, "the necessity of maintaining strict discipline in the various classes."[17] Whether this reflects on Scott's assessment of Dobson's disciplinary strengths, or indicates Scott's own excessive anxiety that "his" School would survive his absence without strain, or again whether it suggests that the local apprentices attended evening classes somewhat unwillingly, there is no way of knowing. It appears, however, that the year ran smoothly in Scott's absence.

Scott's visit to the United Kingdom and Europe was immensely productive. Dr. Pagenstecker, the famous German occulist whom he had consulted at Wiesbaden, was able to reassure him that he was not facing possible blindness. Once this fear had been allayed and his eyes treated, Scott was revitalised. He was able to visit, as promised when he applied to the Board for leave of absence, numerous engineering establishments in Great Britain and Northern Europe. A *Press* report of 3 February 1900, relayed from London provided an impressive description of Scott's activities at this time. He was thoroughly in his element, enjoying the attention accorded him at the various technical schools by a letter of introduction from the Agent-General in London. The reporter had interviewed Scott a few days before he and his wife embarked on the S.S.Oceana for their homeward journey to New Zealand. He wrote:

> Since returning from Wiesbaden, Mr Scott has been flying about in the most remarkable manner, and working tremendously hard, studying all the latest developments of electric traction, electric lighting, railway work, and technical education. He has gone thoroughly over the technical schools at Darmstadt, Manchester, Sheffield, Gainsborough, Leeds and Nottingham, those of the City Guild Institutes, Owens College, King's College (London), Trinity College, (Dublin), Glasgow University, Edinburgh University and Cambridge. He has also inspected the electrical works at Wiesbaden, Frankfurt, Brussels, Darmstadt, Maine, Dublin, Liverpool, Manchester and Glasgow. Also the Clyde shipbuilding yards, and most of the chief railway and manufacturing establishments.

The report concluded with the understatement that this was "a tolerably good programme"[18] for a man who was virtually an invalid on his arrival. It is obvious that Scott's energy and enthusiasm had completely revived. He had already written from Europe requesting permission to spend £50 on equipment, which was granted readily. Certainly the

Board of Governors was to find him a force to be reckoned with when he returned brimful of new ideas and projects in March of 1890.

Edward Dobson

Arthur Dudley Dobson

Though he is seated, Scott dominates the School staff in this 1902 photograph. Standing from left are A.R. Craddock (lecturer in electrical engineering), F. Innes (student assistant), T. Hill (laboratory attendant), E. Cull (demonstrator) and A.F. Morrison (demonstrator).

CHAPTER FOUR

"A Separate Institution"

"The School of Engineering [is] to all intents and purposes . . . a separate institution!"[1] So declared the *Lyttleton Times* in 1900, and the reasons why this should be so are not hard to find. There was, first of all, the separate endowment which meant that the affairs of the School, both financially and administratively, came under a special sub-committee of the Board, the Museum and Library Committee. The School's finances, apart from students' fees and the limited amount coming from testing, came out of the endowment which provided for a Museum, Library and School of Technical Science. G.W.Russell, a new appointment to the Board in 1897, commented on the immense amount of correspondence relating to the concerns of the School of Engineering. He complained too that while he had thought he had been elected a member of the Board itself, he found, on the contrary, that he was merely a member of the Museum and Library Committee.

Once there was provision for matriculated students to follow a University course, the Board felt it was within its rights to nominate a small sum each year to the School from the endowment which provided for higher education within the province, although this was regarded with some suspicion and there were snide comments about "robbing Peter to pay Paul."[2] The School was required to pay the College a nominal rent for the land on which the engineering laboratory was erected, as well as contribute money towards the cost of the teaching in mathematics, physics, chemistry and geology which the College provided. Later the Government allowed a capitation grant towards the cost of educating the apprentices and evening students, and still later a specialisation grant as increasingly it was recognised that the School served a wider area than just Canterbury and that therefore Canterbury should not be required to bear the whole financial burden.

All these economic matters tended to separate the School from the ordinary day-to-day running of the College, and helped to substantiate the uneasiness with which certain members of the College community regarded the incursion into their domain of a professional school, catering for students who wore "dungarees and sandshoes."[3] Such a school sat most uneasily with the traditional view, inherited from the nineteenth century, that a University provided liberal education befitting a gentleman. The cheerful enthusiasm of a writer in the *Press* in 1899 that the new engineering laboratory meant that the engineering students could get themselves "as greasy as they like"[4] contained an element of condescension. Even today despite all the efforts to upgrade the profession by IPENZ the term "engineer" remains ambiguous. To describe a man or a woman as an engineer does not adequately discriminate whether the person in question is a mechanic, a foreman on a factory floor or a professional engineer.

Another reason for the undoubted separatism of the School from the rest of the College was Scott's overweening personality. He certainly tended to emphasise the ways in which the organisation of the School kept it separate from the College, not to downplay them. Thus in 1903, he reminded the Board members that "The School of Engineering, although advantageously associated with the College proper [is] practically a separate institution, of which the Professor in charge is, under the Board of Governors, the sole head and administrator."[5] Scott made much of the fact that he communicated directly with the Board of Governors, not through the Professorial Board and that he submitted an annual report which was then incorporated in the Chairman's annual report of the College. His pride in his separate status and his insistence on protocol helped to intensify the divisions between Scott and the rest of the College. He and Professor Cook evidently had a good working relationship, but it was a different case when it came to Bickerton, another founding professor, who was responsible for teaching chemistry and physics in the Engineering Intermediate year. Scott was seriously troubled by the very poor pass rate his students were achieving under Bickerton's tutelage, and it is widely believed that he played a considerable part in the witch hunt which led to Bickerton's dismissal.

Because of the different status of the School, Scott was able to claim certain privileges denied to other professors. At a time when the Professorial Board was struggling in vain to persuade the Board of

Governors to accept a professorial representative on its august body, Scott was regularly attending meetings of the Museum and Library Committee which dealt with the affairs of the School of Engineering. The spurious reason given by the Board for its obstructiveness towards its professorial staff was that it would place a professor in a most invidious position if he were expected to vote on matters to do with salaries. Thus the Professorial Board had only advisory and not executive status. How impotent it was became very clear in 1906, when several leading engineering students were discovered in the College, celebrating with drink and noisy song after an Engineering Society dinner. The Professorial Board, exercising to the full its powers of disciplining the students, passed very severe sentences of rustication on the offenders. To make matters worse, it had been engineering students, and some of the same ones, who had been responsible for damaging a College ceiling the previous year when they had tinkered with the College clock during Diploma Day celebrations. Board member, L.B. Wood,* seemingly always on the look-out to undermine Scott's position, and appalled at the "disgraceful"[6] behaviour of the engineering students gave notice of a motion that a sub-committee should be set up to examine the situation thoroughly, but he had no supporters. Professor Scott, claiming that his students were "rowdy in spirit but absolutely innocent of all motive of wrong-doing,"[7] was outraged that the students had been dismissed seemingly out of hand. He met privately with the Board of Governors. Afterwards the Board conferred, and following a long drawn-out meeting announced that the sentences had been commuted. Although several of the students were barred from attending lectures or sitting for examinations, this lasted only to the end of that year. The Professorial Board not surprisingly felt it had received scurvy treatment. Its memorandum read: "the Board of Governors, by so materially lessening these punishments . . . has largely nullified the efforts made by the Professorial Board, and seriously weakened its power of maintaining discipline."[8] Its protest met with little joy; indeed the Professorial Board could regard itself as having collectively been rapped over the knuckles. The repressive reply made it clear that it would be wise not to push the Board of Governors too far. It was requested to "refer to the Act, authority, rule or regulation upon which [its] opinion [is] based."[9] This episode clearly illustrates that Scott's position was by then virtually

*Wood, Lawrence Berry 1855-1934; b Scotland; educ Univ of Edinburgh (MA); arr NZ, 1880; teacher, inspector N Canty Ed Bd, 1883-1906; m BG, 1904-34.

unassailable. Wood might complain that the Board was pusillanimous in its dealing with the professor in charge of the Engineering School, but his prestige, public image and the forcefulness of his personality were difficult to withstand.

Other ways in which the engineering course differed from the more usual University courses were that they were more demanding of time, with students having to attend the long laboratory classes as well as lectures; engineering students were castigated in the *College Review* for being "swots,"[10] although the *Press* took umbrage at the term. However it was certainly true that the course was such as to put a great deal of pressure on the students. (This may well explain why it seemed to be chiefly engineering students in those early years who attracted adverse attention by their student pranks; they had to let off steam somehow. Maybe, though, they were just more given to practical jests; today's responsible engineers may well wince when reminded of how they drove a flock of sheep into the south quadrangle, or suspended a bicycle from a college steeple, or poured detergent into the huge water tanks of the Hydraulics Department.)

As well, after the intermediate year, the remainder of the engineering degree kept the students hard at work, enclosed within the precincts of the engineering wing of the College. Inevitably, a professional course is more self-inclusive and inward looking than the more arbitrary and varied courses followed by students in the Arts or Science Faculties, where a specialism may not be declared until the third or even the honours year. Again, the requirement of 18 months' workshop practice before the degree could be conferred created a difference between engineering students and the rest. (The *College Review* made much of the workshop requirement; it offered dire warnings to engineering students as to how to comport themselves appropriately at the Addington Railways Workshops. Apparently the permanent staff took the very existence of the students as an opportunity for playing practical jokes, sending them off to collect such innocent items as strap oil, leather hammers, and striped paint. Most important, however, was the protocol which insisted that "Addingtonese"[11] should be confined to the workshops and not become a regular way of speech.)

One final factor deserves mention: the close relationship which existed right from its very first years between the School and local industry. If there was a measure of suspicion, even hostility, between the School and the College as a whole, the School still looked to industry as its centre. The constant interaction, and the intense co-operation, of which we have glimpsed a little in the encouragement and support offered the School by J.L.Scott, Arthur Dudley Dobson and others, meant that the loyalty of the School was turned outwards towards industry rather than inwards towards the College and University traditions. Any criticism of the value or expense of the School could be countered by a reference to how much Canterbury, and in the end, New Zealand, was benefiting from the work of the School. What was important was that industry and leading practising engineers should approve of the School's aims and achievements; accordingly it mattered less that the College should understand and approve.

CHAPTER FIVE

Electrical Laboratory

Scott's years as professor in charge of the School of Engineering were characterised by successive peaks of high activity while he battled on one or other front to get "more" for his School; more, whether money, equipment, buildings, students or recognition. From each successive battle he would emerge triumphant, "stimulated to fresh effort."[1] After settling down for a year or two, consolidating his gains and basking in the approval won by his success, he would recover his energy and be off again. On his return from Europe, it became immediately obvious what challenge he had set himself: the establishment of an Electrical Engineering Laboratory. His annual report to the Board in March 1900, under a heading ELECTRICAL ENGINEERING, stated without preamble:

> I venture to again bring under the notice of the Board the fact that the teaching of Applied Electricity cannot be omitted from a modern system of Engineering instruction. So generally is this now recognised, that in not one of the 20 Universities, Engineering Colleges, and Technical Schools, which I visited during the past year, is the subject not given special prominence.[2]

He was convinced that it would be possible to run such a department for no more than £225 a year. He suggested that student numbers would also rise, which would swell the income from student fees. They might even reach 200 within five years. Whereas, on the other occasions when Scott had put forward the idea of establishing an Electrical Engineering Laboratory, the Board had shied off the notion very quickly, this time, no doubt because of the prominence given to the need for hydro-electric developments throughout New Zealand, the Board was immediately co-operative. ("In the latter part of the nineteenth century the Mines

Department had been instrumental in keeping the latest advances in hydro-electricity before the public and the government."[3] After 1903 and the passing of the Water Power Act, the Public Works Department concerned itself with hydro-electric development.) The Finance and Estates Committee was accordingly instructed to deliberate on the question of providing plant and accommodation for the teaching of Electricity as recommended by Professor Scott. By 29 May 1900, its favourable report was presented: funds were available for a lectureship at £300 a year; for the purchase of electrical plant as specified by Professor Scott, £1113; for increases in salaries in the Engineering Department of £175; there was even an allowance of £70 towards passage money for the new lecturer; and most important, £1,000 for the erection of a building for the electrical laboratory. However, the Board had obviously not waited for a favourable outcome before advertising the position of lecturer in electricity as a telegram arrived the next day, 30 May, from the Agent-General in London confirming that A.R.Craddock, a former science graduate of the College, who was currently working in Great Britain, had accepted the position. The Board was wholehearted in its support for Scott's project but, understandably, somewhat alarmed at the mounting expenses. It convened a Special Committee which recommended that the Canterbury members of parliament, among them Scott's father-in-law, the Hon. C.C.Bowen, should canvass what support they could muster and approach the Prime Minister, Richard Seddon. The Christchurch newspapers expressed strong approval of both the Board's decisions: to establish electrical engineering and to seek Government support for the project. The *Press* claimed that "the Board are (sic) engaged in a work of colonial importance which should receive colonial support."[4] As planned, a deputation of Canterbury members of parliament met Seddon in October 1900. The Prime Minister was obviously sympathetic but could not promise full support until the passing of the Manual and Technical Instruction Act. This act was passed the same year, and not long afterwards a telegram from Seddon announced that the Government would assist the School's new development to the tune of £2,000.

Justifiably confident that such Government support would be forthcoming, plans had gone ahead merrily for the new Electrical Laboratory. On Scott's recommendation, the name of the School was changed from the School of Engineering and Technical Science to the School of Engineering, Electricity and Technical Science. (At that time

Engineering students outside the School on Diploma Day, 1905.

the role of the electrical engineer was far from clearly distinguished from that of the mechanical engineer, the *Press* claiming, for example, that an electrical engineer was "a mechanical engineer as well as an electrician."[5] Certainly with the bias towards mechanical engineering which was so much in evidence in the early days of the teaching of engineering at Canterbury, it is not surprising that this was how an electrical engineer was regarded.)

In November 1900, Scott was authorised to sail to Sydney by the S.S.Mokoia where he was to obtain equipment for the new electricity department. By 7 January, he had mailed off a confidential letter setting out the advantageous terms he had negotiated with the Sydney engineering firm of Messrs Noyes Brothers, the agents of the Westinghouse Company of America. The equipment, which he claimed he had bought at "roughly one half of its commercial value in New Zealand,"[6] included "a complete Experimental Lighting Plant," which consisted of "a 10 h.p. Generator, a Battery of 30 Storage Cells, [and] a three-panel Italian marble Switch-board, with all necessary switches and instruments." As well, there was "an Experimental Continuous Current Power Transmission Plant," with "a 7 h.p. Experimental Dynamo, a 2½ h.p. Series-wound Motor, a 2½ h.p. Shunt-wound Motor, starting and regulating resistances and switches; [and] an Experimental Alternating Current Power Transmission Plant," which consisted of "a 7½ h.p. Experimental 1, 2, or 3-phase Generator, a 3 h.p. 2-phase-Motor, a 3 h.p. 3-phase Motor, three transformers, 1, 2, and 3-phase, transforming up to 1000, 2000, or 42000 volts." As well as sundry smaller items, he had purchased "a 13½ h.p. Gas Engine."[7]

If the Board looked to this as a possible saving, it was to be disappointed. Inevitably Scott spent the money he had thereby saved on more equipment, which necessitated further costs for shipping, insurance, and installation, even if the Customs Department passed it duty free. No doubt Scott retained a memory of the risk of subsidence in connection with the erection of the testing machine in 1895 when he offered to personally undertake supervision for the foundations for the new apparatus. It would be hard to imagine him standing aside; as the Chairman of the Board was to say in one of the opening addresses of the new laboratory: Professor Scott "had thrown himself so thoroughly into his work that it was to him a work of love . . . "[8] The *Press* acclaimed the splendid new equipment. "To render [the newly imported] plant as complete as possible, the Professor has decided to have an experimental light installation of sufficient capacity to light the laboratories under his control. This plant will be complete in itself, and driven by a gas engine, and arrangements will be made for the students in electrical engineering to be solely responsible for the working of the plant. In this way they will get a thorough insight into the practical working of electrical installation."[9] (Not everyone, however, was as enthusiastic. In June of that year the *College Review* declared that "Professor Scott's new gas engine is not giving satisfaction. It is doubtful whether it was not originally built as a Pom Pom, being very disturbing to lectures overhead."[10] Placing the engine in a glass case was apparently not sufficient to silence it but "the tardy application of a muffler has completely remedied [its] painful asthmatic gaspings . . . and peace and tranquillity reign once more.")[11] But the real advantage was that "the possession of a School of Electrical Engineering would put Canterbury far ahead of other provincial districts and would make their College in that respect a centre for the whole of New Zealand."[12] Thus it was unique to Canterbury. The rather cavalier attitude towards the School in the early years was slowly shifting ground so that, instead of having to fight every inch of the way to get the equipment and buildings which he regarded as necessary, Scott could be confident that his demands would be met. Initially, the Board had grudged every penny it was forced to expend on engineering plant but now the realisation had come that it was greatly to Canterbury's advantage to have a special school, which attracted students, prestige and Government money. Obviously, there were still a few dissenting voices. In August 1902, Scott was again rebutting the charge that the College was impoverished by the School of Engineering. By and large, however, it was increasingly accepted that the School was a necessary and valuable adjunct of the College, and worth the expense.

Craddock, the newly appointed lecturer, chose to return to New Zealand by way of the United States in order that he might learn as much as possible about engineering education in that country. Shortly after his arrival at the beginning of 1901, it was announced that he would deliver a series of public lectures on "Electricity and its Applications". However, although the new member of staff, the new equipment from Australia, and news of the Government specialisation grant were all on time, the necessary building was not to be ready until part way into the first term of the following year. If electrical engineering was to be taught, then a building would have to be found. Undaunted, Scott requisitioned the College Hall, first promising faithfully that it should suffer no damage in being converted into an electrical laboratory. The *Press* thought that it presented a very "business-like appearance . . . with one half temporarily partioned off as a lecture room for Mr Craddock . . . the other portion being occupied by Professor Scott . . . who has a large amount of apparatus erected for experimental purposes."[13] Evidently the pioneering spirit was not dead; nevertheless it cannot have been easy for Craddock to begin his lecturing career under such difficult circumstances.

Meanwhile the new building designed by the architect, C.J.Mountfort, was proceeding and warranted an enthusiastic description in the *Press* on 23 January 1902:

> The whole is very solidly built, and in the magnetic testing room there is no iron, either inside or outside the building, the floors being fastened with copper nails, while the fittings are all brass or copper. Piers are built up from a solid foundation eight feet below the floors, to ensure absolute rigidity . . . [There is] a fine lecture room and private room for Professor Scott.[14]

The students were more satirical. "That Canterbury College should ever descend to an extension in red brick was, no doubt, unthinkable by its earliest rulers, but such now is its fate . . . " (It was not until 1915 that the offensive red brick was faced with grey stone to match the rest of the College buildings.) The new Electrical Laboratory, was "of an aristocratic quality, and unadorned beyond the verge of plainness, [bearing] on its face the impress of the age of utility." Not unnaturally they complained about the lack of a water supply in the building, and found "the new lecture room . . . just as inconvenient as the old — it has just the same ingenious contrivances for preventing a good view being obtained, either of the blackboard, or of the experimental work on the Lecture Bench."[15]

The building was still not completed for the first term of 1902; as a result the official opening was timed to coincide with Diploma Day. The speakers, apart from the Chairman of the Board, were Sir James Hector, Chancellor of the University of New Zealand, and Scott himself. The opening was a great bonanza with afternoon tea at 1/- a head, although the price was to be dropped to 10d if there were more than 100 takers. The School was also open to the general public in a conversazione so that the citizens of Christchurch could admire the new apparatus. Scott delivered a speech on the importance of the engineering profession during the Diploma Day ceremony as well as at the actual opening of the new building. He had already been showered with compliments and he claimed that "he felt quite taken aback by the, to a certain extent, unexpected compliments."[16] The Chairman had read a letter from Mr T.Ronayne, General Manager of the New Zealand Government Railways, apologising for his inability to be present, and applauding the Board of Governors for this exciting new venture into electrical engineering. In his opinion, "Canterbury College is singularly fortunate in having Professor Scott as the head of the school, possessing, as he does, a thorough practical and theoretical knowledge, without a happy combination of both no engineer can expect to make headway in the profession."[17] Sir James Hector described Scott as "a man in a thousand"[18] amid applause. The students had already signified their attitude towards Scott's great day by vigorously singing a "Dies Irae, Dies Scott", a risky gesture considering Scott's rather humourless sense of his own dignity. In his brief address, Scott claimed that "at no time . . . had the influence of the University, or the well-being of the community, been greater than at present". What had once been " 'rule of thumb' operations" had been refined "into scientific processes. If a country was to hold her own in the markets of the world, she must now call in the assistance of the chemist, the physicist, and the engineer to aid and direct progress in her agricultural, mining, and manufacturing industries . . . The appliances of electricity, the study of which involved the highest mathematical and physical knowledge, had extended into every walk of life." He acknowledged "the extraordinary clearness of vision" of the early colonists who had not held the mistaken notion that what was needed was "trades' education"; no, "they had seen that the true function of a University College in this direction was the training of those who were to be the brain workers of industrial organisation." In the light of the confusion of motives — pragmatism, idealism and provincial rivalry — which were rife at the time of the School's establishment, this retrospective self-congratulation is somewhat derisory. But Scott was fairly launched. He believed so totally in what he was doing and that "industry today must be controlled and directed not by the so-called 'practical' man, dependent on the limited experience of a lifetime, but by the educated mind trained to so direct scientific knowledge as to satisfy the practical wants of mankind."[19] Implicit in Scott's definition of an engineer is the ringing phrase which emerged at the time of the founding of the British Institution of Civil Engineers,* as well as his attempt to demarcate the limits of professionalism in engineering.

In his second address on this important day, Scott graciously acknowledged the presence in the audience of Edward Dobson, by this time 86 years old, and declared that the School could not be regarded as functioning adequately while it lacked any teaching in civil engineering. (Indeed, this was to be the next challenge he was to tackle.) Amid all this acclaim and approval, however, it was inevitable that a sour note should be sounded somewhere. It was provided by the *Press*, which, while admitting the achievements of the School, still felt that a cautious warning was appropriate:

> Not only can the College now boast of well-equipped engineering and electrical laboratories, equal, as Sir James Hector tells us, to anything in Australia, in efficiency, if not in point of size, but as the facts quoted by Professor Scott show, students qualified at Canterbury College have succeeded in obtaining important appointments all over the world. We have little doubt that with the impetus given to it by the opening of the new building the Engineering School will go on progressing, and that it will attract students from all parts of New Zealand. It would be a pity, we think, if the "boom" in engineering were to lead students not really "cut out" for the profession to join in the rush because it is the fashion.

The article expressed concern that "the arts courses are being neglected" to the possible future detriment of society. "We are great believers in the arts course, if only as supplying an element of culture which should help to make good and capable citizens and add refinement to their lives."[20] The nineteenth century belief that aesthetic sensibility indicated moral refinement had not been eroded by the devastation to be wrought in Europe by two world wars. This is not to say, however, that there has not been continuing concern within Engineering Schools, especially in the States in recent years, that the courses they offer are too narrow and could do with a leaven of the humanities, but with increasing specialisation it is becoming harder to intrude arts subjects

*"The art of directing the great sources of power in Nature for the use and convenience of man." Thomas Tredgold, 4 January 1828.

into the already very pressured engineering courses. By the same token, it could be said that an arts student might well benefit from a technological course or some exposure to scientific method. Anyway, it is a perennial problem and it was well that the writer for the *Press* should alert the general public to possible disadvantages. In the flurry surrounding the opening of the electrical laboratory, however, the *Press*'s cavil did not attract much attention. The splendid new equipment was displayed to the citizens of Christchurch in a conversazione and was "a decided success."[21] It was only later that dismay was expressed that the evening students did not get the chance to do anything more than look at the new equipment.

Although Craddock was obviously very happy to return to New Zealand, he only lasted at the School three years. After an acrimonious exchange between him and Scott, he wrote that he was not seeking re-appointment; however Scott had actually pre-empted him by informing the Board that he could not recommend that Craddock's position should be renewed. From the itemised list of grievances, which Craddock sent to the Board of Governors, it is possible to gain some insight not only into Scott's autocratic manner of directing the School but into its day-to-day routine. And while Craddock may well have had reason to regard Scott's attention to detail as tiresome and footling, it was that very attention to detail which ensured the smooth running of the School. At various times during the three years he was teaching electrical engineering and electricity, Craddock received memoranda in which Scott stated very precisely how he wanted certain routine matters dealt with. He issued instructions as to how Craddock was to keep the Storage Battery Log Book up-to-date, how he was to record the practical work of students in the various classes, when he was to take his vacation and what the borrowing procedure at the engineering library was. The following memorandum for Mr Craddock, dated 22 June 1903 and signed Robt.J.Scott, Prof.in-Charge provides a clear impression of how Scott ran his department:

> It will be necessary for you next term to give instruction in drawing and designing Electrical Machinery to a class which will meet from 2 to 4 p.m. on Monday, Wednesday and Friday.
>
> A standard set of designs will be settled on.
>
> The outline of these designs together with a summary of the steps

> in, and results of the calculations will be entered by you in a book which will be provided for this purpose.
>
> For the work of next term you will please prepare and submit to me at as early a date as possible an outline design with calculations for a C.C.Multipolar Compound wound Generator* of about 30 K W capacity.

Craddock found some of these instructions "arbitrary" and resented the "peremptory" tone in which they were delivered. Other complaints suggest that there was a real communication problem between the two men, in that he felt that there were "unnecessary obstacles placed in the way of [his] carrying out [his] duties."[22] Scott, on the other hand, describing Craddock's behaviour as "passive obstruction,"[23] was puzzled as to why the latter had not spoken to him about his difficulties. Craddock obviously felt blocked on all sides by Scott's supervision, which certainly did not encourage initiative or creativity in a subordinate. And one of his major grievances was that he had assumed that, as a lecturer, he would automatically have a seat on the Professorial Board and some sort of standing in the College. Certainly the *College Review* described him as "in charge" of the Electrical Laboratory; Scott, however, was outraged at such a suggestion. " . . . had he a seat on the Professorial Board others engaged in lecturing at the School of Engineering would be entitled to a similar privilege, which would result in the possibility of a block Engineering vote being given against the interests of the College proper, or of my being out-voted by my own staff, conditions which neither the Professorial Board nor myself could be expected to subscribe to."[24] A clearer statement of the role of the god-professor would be hard to find. In the terms of appointment drawn up at the time of the appointment of Craddock's successor, there is the unequivocal statement that this position did not carry with it a seat on the Professorial Board.

It would seem that Craddock was immensely ambitious and found his subordinate position, "under the supervision of the professor in charge" most unpalatable. He took the unprecedented step of submitting a list of his grievances directly to the Board of Governors; Scott's response to this attack on his authority was a 20 page letter, in Morrison's handwriting. Craddock had been appointed on Scott's recommendation; apparently some members of the Board had been troubled at his "youth

*Continuous Current, see p. 61

and inexperience" but Scott had offered the reassurance that he "would give personal attention to the Electrical Engineering portion." Thus, he asserted that "I have arranged the courses and exercised supervision in the Laboratories especially with regard to Electrical Engineering, and during the first year of Mr Craddock's term, when, owing to his inexperience in lecturing, complaints were frequent, and students steadily falling off, it was only by my intervention, support, and advice, that a collapse was averted." If Craddock's ability to keep discipline was suspect, it was certainly the case that the matriculated students appreciated his willingness to give them assistance outside regular teaching hours.

The Museum and Library Committee, faced with Craddock's grievances and Scott's rebuttal arranged a special sub-committee to interview both men. But by then, Scott had made his position completely clear: "it is impossible for the work of a Department, in which discipline must of necessity be a first consideration, to be carried out with a member of the staff in opposition to the Head."[25] This was July. A public announcement that Craddock was not seeking re-appointment appeared on 1 September.

Scott's workaholic streak meant that his expectations both of himself and of his staff were excessive. Craddock's complaint that he found it unreasonable that he was expected to produce such large quantities of diagrams to illustrate his lectures was greeted with the rejoinder: "It is true that the Engineering Department possesses a large number of diagrams. Of these a very large proportion was prepared by myself when carrying on the whole work of the Department singlehanded. A similar course was open to Mr Craddock . . . "[26]

Once Craddock had officially resigned, the final term of that year, 1903, must have been very fraught and uneasy. Scott's eye troubles were flaring again, and he was forced to request a further year's leave of absence at the beginning of 1904. The Board granted him the year, once more on full pay, and he made another visit to England and Europe. While Scott's yachting pursuits would have given him some respite from the pressures of work, obviously he drove himself to the limits of his strength. A medical statement about his health which he forwarded immediately after his arrival in Great Britain indicated that his iritis had returned because he was "run down and his system out of order". If he paid "careful attention to diet and exercise"[27] he would be able to prevent another crippling attack. But, by the beginning of 1904, with his eyes becoming worse, he was understandably very despondent about the School and the teaching of electricity. A letter in Mrs Scott's hand of 19 January 1904, revealed that he was "not sanguine"[28] of finding a suitable man to replace Craddock in the colonies. Inevitably there had to be makeshift arrangements to fill the gap left by Craddock's resignation. L.Birks, later Chief Electrical Engineer for the Christchurch M.E.D., accepted a temporary position but had to be held to his word when a better career opportunity presented itself. At the same time, C.C.Farr* was appointed part-time lecturer in electricity.

The position was once more advertised overseas, and Scott, while he was in England was able to interview, and subsequently appoint P.H.Powell to the lectureship. Powell had an M.Sc., as well as an M.Eng with first class honours from Liverpool University. After three years' research work, he had joined the electrical firm of Siemens Brothers, Stafford. "His knowledge of theoretical electrical engineering was without parallel . . . "[29] This appointment was to prove most satisfactory. There is no evidence of friction — as the result of misplaced ambition — between Powell and his superior. On the occasions when he applied for leave of absence over the long vacation in order that he might visit relatives in the United Kingdom, Scott was actively supportive. Like other members of the School, both staff and students, Powell was also accorded the privilege of sailing with Scott. A letter, headed *Yvonne*, Pigeon Bay, was sent on 2 February 1911 to the Registrar. After dealing with engineering business, it concluded cheerfully: "It is a beastly day here blowing and raining from the South. I have just put Powell who is with me ashore. We got over the bar yesterday very comfortably."[30] It does not appear that Powell became a devotee, although of Scott's later staff members, J.E.L. Cull** and Percy Bamford, both of them graduates of the School, were keen yachtsmen. There was an inevitable delay before Powell could take up his appointment in 1905. At this stage in the School's development, the staff consisted of Professor Scott and two demonstrators, Morrison and Cull, (who had replaced Otto

*Farr, Clinton Coleridge 1866-1943; Physicist, researcher, teacher. B Adelaide; educ St Peter's Coll, Adelaide Univ, D Sc; lect Sydney Univ, 1891-5; undertook magnetic survey NZ, 1898-1904; established Magnetic Observatory ChCh; lect physics and surveying, Sch of Eng CC, 1904; lect physics, 1905; prof, 1911-36; FRS FPhSL, FNZ Inst, Hector Medallist, 1922. Farr was a leading scientific figure in Australia and NZ for 30 years.

**Cull, John Ernest Lelliot 1879-1943; B ChCh, educ CBHS, CC B.Sc(Eng); demonstrator Sch of Engineering, 1904-8; employed on designs and in charge of construction for the Auckland main drainage scheme, 1911-2; Designing Engineer, P.W.D., 1914-30; prof Civil Eng, CC, 1930-1942; Cull did much pioneering work in the investigation of the smelting of Taranaki iron sand by means of the electric furnace.

Peez in 1902 when the latter had felt that he was being paid an insufficent amount for increased work.) Otherwise there was a succession of student asistants; both of the later professors R.A.Campbell* and S.Steele started their long association with the School at this time by serving in this capacity.

The episode with Craddock highlights the problem, not confined to the Engineering School, of attracting and keeping staff below the level of professor. In other departments of the College, what has been called a "cultural cringe" operated, in that while chairs were filled from overseas, the lecturers tended to be homegrown products. The School of Engineering, over the years, retained a policy of appointing old graduates. But in these early days, with only one engineering school, an applicant for a position in the School, if he were a New Zealander, would inevitably have already passed through the School as a student. And, the position of demonstrator was a thankless one, a dead-end with no assured prospect of either an increase in salary or a rise in status. On the whole, men took the position on a temporary basis, remaining, if the School were lucky, for half a dozen years, before leaving to advance their practical experience. Scott always specified at the time of a new appointment that ideally he required a man with several years' practical experience but more often was forced to accept a recent graduate with no outside experience. He was obviously most fortunate in gaining a man of Powell's eminent qualifications, but then the position was that of a lecturer-demonstrator, not the more menial position of demonstrator.

---

*Campbell, Robert Archibald (1881-1955). Educ CC, B.Sc.(Eng), 1902; civil engineer on the Bengal and North-western Railway, 1908-12; P.W.D., N.S.W., 1912-19; Campbell joined the Australian Imperial Forces and served in the Palestine campaign as a lieutenant of engineers; lect civil engineering, CC, 1920, and prof 1923-9; consulting engineer Campbell and Morrison, (later Campbell and Hamann). CC Council 1933-6; 1942-8.

CHAPTER SIX

Civil Engineering

Always an opportunist, Scott in his address at the time of the opening of the Electrical Laboratory in 1902, had not missed the chance to refer to the need to reinstate civil engineering in the courses offered by the School. It is not surprising, therefore, that his next forward drive was the establishing of a Hydraulics Laboratory. Once Edward Dobson's services had been dispensed with in 1902, the courses which he had taught in surveying and the principles of civil engineering had disappeared from the School, although civil engineering was still offered on the syllabus. While all Scott's attention was directed towards establishing the mechanical, and later the electrical engineering laboratories, civil engineering was low on his list of priorities. But now, he regarded this as his next major challenge. The old rallying cry of provincial rivalry was as effective as ever. On 28 August 1903, Scott sent a confidential letter to the Board with his proposals for the School of Engineering for the following year. The need for civil engineering was outlined — it appears there had been several applicants — and then came the crunch: "There is a considerable amount of agitation in the Wellington Province for the setting up of a School in Civil Engineering at the Victoria University it being argued that this branch of Engineering is not provided for here, and as long as this argument exists there is the danger of outside pressure forcing the University authorities to establish such a school although it is generally admitted by them that a large portion of the Engineering course being common to all branches such duplication must be injurious to both Institutions." Obviously, however, it would be expensive to set up civil engineering. ". . . an increase of £750 per annum in income will be required if the progress of the School is to be unchecked." Scott fully realised that he must tread warily. "I cannot, in fairness," he acknowledged, "ask the Board to vote such an amount from the parent funds for work which is fast becoming more National than Provincial in character."[1] When 40 per cent of the matriculated students came from outside Canterbury, even if the numbers at this stage in the School's development were very small, the School's claim for an annual specialisation grant seemed unassailable. The Museum and Library Committee agreed that a deputation, led by Scott, should again petition Seddon for financial assistance. Meanwhile, Scott was battling with the loss of his full-time lecturer in electricity and the deterioration in his eyes. Nothing could be achieved in a hurry, but on his recommendation Dr Farr, already teaching part-time in electricity, was offered a part-time position in surveying for 1904.

The soliciting of the Premier again proved effective and an annual specialisation grant of £1,500 was promised in March of 1905. The following year, it was increased to £2,000. Inevitably there were strings attached. A private meeting between Scott and the Secretary of Education had revealed that it was most important that the grant not be used "to indirectly relieve other departments."[2] Scott was perfectly happy with such an arrangement, but it was conditions such as these which gave rise to resentments in other and struggling College departments. A persistent belief lingered that the College as a whole suffered in order that the School of Engineering could flourish. The School of Engineering could well riposte that the College would not have had access to that money no matter how pressing its claims, but when there was a shortage of government funding for education in general, a large amount siphoned off in one direction inevitably meant that there was less for other areas. University education in New Zealand throughout the years suffered greatly from inadequate funding, until by 1931 it was regarded as "the bag of boiled lollies"[3] offered as a sweetener in the absence of more substantial assistance.

Ernest Rutherford, visiting New Zealand in 1905, referred to Professor Scott's "untiring industry."[4] With an opportunity to develop a new department, Scott was unstoppable. He claimed that the Government grant of £1,500 rendered further extensions of the School necessary — an encroachment on to the Boys' High School playground — and immediately Scott was applying to the Board to contribute a further £1,600 and submitting plans for a more ambitious hydraulic laboratory. In marked contrast to the earlier cautious parsimony exhibited by the Board, now in 1905, the Hon. G.W. Russell could declare: "they must realise that in connection with the School of Engineering . . . a cheeseparing policy was impossible. They desired to make the School of Engineering what he believed it could become, the best institution of its kind in the Southern Hemisphere, and they must keep on spending."[5]

Meanwhile, it was necessary to provide lectures on various aspects of civil engineering; a part-time lecturer in surveying was insufficient for Scott's purposes. He, therefore, recommended the Board to invite Cyrus Williams* to give a course of 26 lectures on Building and Construction. This proved a satisfactory measure and the following year, 1907, the policy was extended; as well as Cyrus Williams, George Phipps Williams** gave eight lectures on borough engineering and Arthur Dudley Dobson eight on railway engineering. The professional engineers co-opted by Scott were already associated with the School by their serving on the committee of the Engineering Society, which Scott had established in 1897, and participating actively in its projects. Thus, Cyrus Williams and Arthur Dudley Dobson, had both accompanied members of the Society on 28 June 1904 on an excursion to the Waimakariri River when there was a possibility that hydro-electric works might be developed there. J.L. Scott's association with the School never wavered. Apart from his undertaking to read Scott's Theory of Workshops lectures in 1904 during the professor's absence he had, by this time, been elected to the Board of Governors. He even felt it incumbent on him to make the public statement that "he would take care to criticise [the new course] if he felt it needed criticism."[6] He did not appear to have found this necessary. The Board requested the two Scotts to confer over the most suitable way of spending the £1,225 equipment allowance; one month later, in October 1907, they had apportioned the money so that £1,000 could be spent locally and the remaining £225 on equipment which would have to be imported.

Progress on the new Hydraulics Laboratory was very slow. When W.R.Robinson, a licensed surveyor, — the only applicant — was appointed full-time lecturer in surveying in 1910, he had to take up quarters in the ground floor of the old Chemistry laboratory. Equipment designed by Scott was on display for a Diploma Day conversazione

*Williams Cyrus John Richard, 1862-1942; B Sydney; educ Sydney and Brisbane Grammar Schools; trained as civil engineer specialising in harbour works under W.D.Nisbet; Engineer Lyttelton Harbour Board, 1902-27; pres NZ Soc of Civil Engineers, 1920-1.

**Williams George Phipps, 1847-1909; B London; educ Cambridge, BA; trained under Sir Joseph Bazalgate; arr NZ 1869; Resident Engineer for ChCh-Rakaia rlwy; involved in other rlwy works, the Lyttelton Harbour and water race irrigation schemes; Engineer and Sec to Waimakariri and Ashley Water Supply Boards.

in the same year; it was greeted enthusiastically by the *Press* as "the most complete and up-to-date . . . south of the line."[7] However money was short; engineering scholarships were temporarily withdrawn and there were references made to the "depressed state of finance"[8] of the School. Aware of the "unstable conditions of Politics"[9] at this time, the Board proceeded cautiously; it hoped that the Government would oblige with a £ for £ subsidy, which it had previously allowed, but thought it wise to gain an actual promise before authorising the additions to the School. The amount required was £5,500, ie, each party should contribute £2,750. (The comprehensive range of Scott's directorship of the School can be clearly discerned from two consecutive letters which he sent to the Board in 1912. The first requested the £5,500; the second asked for better accommodation for staff bicycles.) The *Press*, very partisan, especially in the light of moves from Auckland towards establishing its own Engineering School, was gratified to see that "the Governors of Canterbury College are still pursuing a progressive policy."[10] The Government, however, was less enthusiastic; a telegram from Wellington read: "Minister regrets that he is unable in view of large commitments on account of school buildings to entertain at present application in aid additions School of Engineering."[11] Scott was outraged. He was rather of the opinion that the Minister had not fully appreciated the situation. "I would . . . suggest that the Minister be again approached and the point at issue be clearly put before him. It is simply that the New University Courses in Engineering cannot be carried on at Canterbury College without the provision of additional accommodation."[12]

Scott's response to the Government's refusal also expressly warned that if the new courses could not be offered the following year, then there was another University College in the country, ready and waiting to take up the slack. If letters failed to convey the seriousness of the situation, then a personal interview was obviously necessary. Accordingly it was decided in October 1912, that Scott and the Chairman of the Board should go to Wellington to personally interview the Minister, George Fowlds. They were successful. Authority was granted and the three-storey extensions, designed by Collins and Harman given the go-ahead. The alterations and additions to the School, completed in the end in 1914, provided an "additional lecture room, a departmental library, workshop, a room for applied mechanics, a private room for a lecturer, and a preparation room."[13] Unfortunately Scott was "incapacitated through illness"[14] and was therefore not around to supervise the construction of these "advantageous surroundings."[15] However, he had apparently instructed his staff to keep a watchful eye on the progress of the building. They all reported anxiously that the piers supporting the hydraulic laboratory needed strengthening if they were to carry the expected load. Scott obviously concurred and wasted no time in warning the Board, who wrote to the architects, Collins and Harman, advising them of Scott's opinion. They were not impressed. "As he is not an Architect we do not consider his opinion or design worth consideration,"[16] they retorted. The Board members could not have relished their invidious position as middlemen between two antagonistic groups of professionals. Scott was undeterred; he and his staff were genuinely concerned that there was risk of a "catastrophe". The Board could not ignore a letter marked URGENT, which warned that "the Main Floor of the New Engineering Building [was] on the verge of collapse by failure of the columns in the Hydraulic Tank Room and that it [might] come down at any moment."[17] Scott was instructed to arrange that the piers be strengthened. This was not the first time that Scott had had a run-in with the College architect. He performed some tests on the Oamaru stone which was used for facing the College and found it unsatisfactory. Far from receiving appreciation for his gratuitous report, it was apparent that he was regarded as meddlesome and interfering.

By the beginning of 1913, Scott's health had deteriorated badly; he had, once more, to apply for leave of absence. His medical certificate stated that he had received gas poisoning which had turned to neuritis; as well, he was suffering from gout. The illness was attributed to "a long continuation of brain work," and a trip overseas was recommended. "His brain at present is quite unequal to the strain of either consecutive thinking or worry."[18] Judging from the shaky quality of his handwriting, Scott was decidedly frail. Leave was granted; only this time, instead of allowing him full pay, the Board insisted that he pay half the cost of a substitute. This was the policy which obtained for other College staff when they were granted leave of absence. Scott complained bitterly; but to no avail. He recovered amazingly during his year overseas and, on his return, gave an address to the College, entitled "An Engineer's Holiday."

> . . . to an engineer, even an engineer on holiday, so many examples

The Engineering laboratory in 1902 with the 50-ton testing machine on the left and the experimental engine at centre right.

of the profession and craft were continually cropping up that it was hard to get away from his job. Personally, he had desired complete rest, but soon found himself involved in the theories of that very sport which he was following up for the purposes of relaxation, to wit, yachting.

As with his two previous visits, he had spent much of his time inspecting engineering colleges throughout Great Britain and Northern Europe, and making comparisons between what they had to offer and the Canterbury School. He was completely satisfied: ". . . nearly all the colleges suffered from the difficulty of providing practical experience as well as theoretical instruction, and few, if any, possessed the advantages of Canterbury College in being able to arrange for the practical work which was necessary for a complete education."[19] With his batteries thus re-charged, he was eager to resume his work with the School.

Concentration on the dramatically eye-catching aspects of the School's history must not obscure awareness of other areas of development. One such development was the establishment by Scott of the Engineering Society in 1897. The officers elected for that first year included Scott himself as President, and practising engineers, John Anderson (of Andersons), A.L.Beattie (a Locomotive Engineer) and J.L.Scott as Vice-Presidents. (This association of non-students with the Society meant that in later years it was consistently refused a percentage of the Students' Association fee which other College Societies could claim.) Scott's first presidential address was published verbatim in the *College Review*, after it had been "urged that the engineering students had as much right to be considered in supplying reading matter as the art students, especially as the disciples of pistons and dynamos outnumber the former."[20] Besides, the editors acknowledged that, despite their expectations, it was not too technical. Scott outlined the possibilities for a Society which served the needs of two different groups of engineering students "being trained by somewhat different methods":

> The bringing into intimate contact by this Society, and the interchange of views upon any subject, between those who must of necessity approach it from such different standpoints, cannot fail to be of the greatest advantage to all. It will show to those who are apt to rely solely on the knowledge of the abstract theory, that a knowledge

> of such theory alone is not the only thing necessary in the carrying out of an engineering undertaking, but must be accompanied by an intimate acquaintance with the methods and details of practical working. To those who take pride in applying to themselves the title of practical men, it will soon become apparent, that when thrown on their own resources, a knowledge of methods is of little utility when the first principles on which these are based are not clearly understood.

In addition to this attempt to reconcile potential differences between the matriculated and "extra" students, Scott offered a comprehensive list of possible subjects for papers which could be read at the Society's meetings. He claimed that they were "practically innumerable." As well, "another very useful function of this Society would be the organising of excursions to engineering and manufacturing establishments, and to engineering works and progress. If it were possible for these excursions to be immediately followed by a spontaneous discussion on what had been seen, their utility would be greatly increased." He wound up his speech by wishing the Society "a long, prosperous and honourable career" and expressed the hope that "when Macaulay's New Zealander survey[ed] the ruins of London bridge he [would] have the good sense to call on the professional assistance of, and as yet unborn members of, the Canterbury College Engineering Society."[21]

In 1898, the *College Review* reported that the meetings had been reasonably well attended and that discussion arising from the papers was good, "although it is certain that many brilliant ideas have escaped publicity from the fact that most of the members are not used to getting on their hind legs."[22] Meetings were originally held fortnightly, but this was found to be too onerous, and it was resolved, in 1900, that henceforth they should be held on a monthly basis. Later, the Society with some slight financial assistance from the Board, was able to publish annually an account of its *Proceedings*.

Although Scott had established a strong relationship with the Board of Governors, who only occasionally found it necessary to rein in the headlong flight of the engineering professor, he still had to endure the galling restrictions imposed by "the legislation of a body, [the Senate of the University of New Zealand], having little acquaintance with the requirements of the engineering profession."[23] Fortunately, a change in the constitution of the Senate made it possible for the Board, in 1903, to elect as its representatives to the Senate both Scott and his father-in-law, the Hon. C.C.Bowen. In this way, Scott was firmly established as the Senate's "mentor and guide in the intricacies of engineering technicalities."[24] This also meant that, as the years went by, Scott was almost invariably convenor of the various committees of the Senate which dealt with engineering or mining matters, very well placed, therefore, to block Auckland's pretensions to an Engineering School.* The year after Scott and Bowen were elected to the Senate, the engineering degree was established in its own right. B.E. (and later, even M.E.,) replaced the earlier B.Sc.(Eng). Scott's next move vis-à-vis the Senate was to suggest in 1906 that it would be a shrewd move if these degrees were to be brought into line with the requirements of the prestigious British Institution of Civil Engineers. Scott had discussed such a possibility with the chairman of the Institution, Sir Alex H.C.Kennedy, when he was on leave in Britain in 1904. As a result, Technical Chemistry appeared on the engineering syllabus and the School had to pay £50 a year to the Chemistry Department for its teaching. By the next year, a jubilant Scott was able to report that he had gained the crucial recognition from the Institution of Civil Engineers.

> Completion of a course in Mechanical, Electrical or Civil Engineering at the School of Engineering, and obtaining the University degree in the subject now exempts the holder from sitting for the Institution's examinations for Associateship Membership.
> At the date of this recognition McGill was the only other University outside of Great Britain on which this honour had been conferred.[25]

This concession was all the more remarkable in that the teaching of civil engineering had only just been re-instated in the B.E. degree. The *Press* proudly declared that this "recognition placed a hallmark on Canterbury College," and referred to the "exclusive company"[26] which the College had now joined. (Just how exclusive was yet to be seen. In later years, Auckland University College was to strenuously attack Canterbury for blocking recognition of the degrees conferred by its School of Engineering. Canterbury never acknowledged that it was culpable of such underhand dealing.) This international recognition of the degrees of the Canterbury School must have been most welcome to Scott; it certainly explains why, among the professors throughout

*see p. 148

New Zealand, he was very slow in pressing for the abolition of the system of external examiners, believing, possibly rightly, that it had influenced the Institution of Civil Engineers to grant its exemption.

For all the public acclaim that Scott was winning, there were times when his behaviour seemed to be unconscionably high-handed. One issue that was brought to public attention was that he had, apparently, refused to free Wednesday afternoons, so that engineering students could attend their all-important football practices. His behaviour was described in an irate letter to the *Press* as arrogant and typically unco-operative. (Engineering students made up the majority of the College football team.) This attack was unjustified as Scott was very sorry that he was unable to release the footballers, sent a most reasonable letter to the Secretary of the Football Club explaining the exigencies of timetabling and at the beginning of the following year, 1906, wrote to the Board that he "had been able to clear Wednesday afternoons."[27] It was perfectly true that once the timetable had been arranged for any one year, it was impossible to disrupt it, especially with the very full programme necessitated by the two-fold function of the School. Actually, by this time, it was virtually three-fold. Concerned at the number of students who were not making the grade because of poor mathematics, Scott, on his return from his first overseas trip in 1900, with the Board's approval, had introduced an associateship certificate for matriculated students who were not quite up to the full degree. As it earned a certificate, and not a degree, it could be examined locally. The associateship, with its less strenuous emphasis on mathematics, occupied a four year span, in comparison with the degree, which, at this time, still took five years including the intermediate year. (At this stage, too, failure in one subject, meant a whole year's delay, as it was impossible to "carry" a subject over to the next year, a very draconian system.) Many good men went through the School at the associateship level, including two later prominent staff members, G.G.Calvert and T.R.Pollard. It was not, however, recognised by the Engineers Registration Board, which was established after the 1924 Engineers Registration Act, and it was finally phased out in 1929.

CHAPTER SEVEN

"Room for Infinite Ructions"

Scott won the majority of his battles with the Board of Governors with the notable exception of his attempt to block the establishment of a Physics Laboratory. That in itself would have been hard for him to endure but the situation was made more galling by the fact that the first professor of physics had started his academic career at Canterbury College as an understrapper of Scott himself. He had been a part-time assistant in surveying — before Robinson's appointment — as well as part-time in electricity. Scott must have had some measure of respect for the man, Coleridge Clinton Farr, as he allowed him to assume administrative control over the School in 1904 while he was overseas. And certainly the year ran smoothly enough in Scott's absence. While Farr was a member of Scott's department — what L.B.Wood described as "an overworked underpaid assistant"[1] — he was obviously accorded the same sort of benevolent paternalism which Scott gave to all his staff provided they exhibited absolute loyalty to him and the School. When Farr was required to repeat some Mathematics classes, Scott was concerned that he was overworked — although a cynic might suggest that the concern was that the engineering teaching would suffer, not that the teacher himself might be excessively strained. He did recognise, however, that a loss of a class would mean a drop in salary. It would appear that there was no personal animosity between the two men until Farr started pushing for the establishment of a separate Physics Laboratory.

At the beginning of the College's history, Physics and Chemistry were taught under the same umbrella by Professor Bickerton. He was, however, erratic in his teaching and it seems very likely that complaints emanating from the School of Engineering about the low pass rates of engineering students taking Bickerton's classes played a major role in the final vendetta against the man who had taught Ernest Rutherford. Just before Bickerton's ignominious dismissal in 1902, the Board, acting on Scott's recommendations, separated Physics into two components; Heat, Light and Sound continued to be offered by the College, while the School of Engineering provided the Electricity and Magnetism. The first moves to establish a separate Physics Department were somewhat tentative but even so, Scott bridled. His objections were, as usual, very strongly worded. On the surface they appeared to be concerned mainly with the quality of the teaching and the possible deleterious effect on the engineering students of a new dispensation. As ever, there was a sub-text. Scott was only too aware of the need for stringent financial caution; it therefore was most unlikely that the Board could or would duplicate the equipment that Scott already possessed for the teaching of Electricity and Magnetism. In that case, either he would have to surrender that equipment or consent to the newly appointed Professor of Physics entering his domain, using his equipment, and, as Scott saw it, usurping his authority. All that was anathema to him. He was accustomed to being "master of his own ship";[2] he required that his staff should be subordinate to him, not his equals. A very long drawn-out Board meeting discussed the dilemma. Bishop Harper, realising that the real problem was that Scott could not endure the thought of sharing authority commented that there was "room for infinite ructions."[3] However, at a further meeting on 22 December 1908, L.B.Wood, astringent as ever, expressed his belief that there was no need for the Board to be protective towards Professor Scott:

> Had Professor Scott not been infatuated with the notion that the engineering school was Canterbury College, he would surely have felt it his bounden duty to promote the progressive policy proposed by the Board in 1899 [the first time a separate chair of Physics was mooted], and instead of the department of physics dragging out a precarious existence during the past decade, it would have been firmly established in a condition of efficiency . . . If the Board could only induce Professor Scott to abandon his dog-in-the-manger attitude and put the resources of his establishment freely at the disposal of the Professor of Physics for laboratory practice, the students of Canterbury College would have better opportunities for practical work than any other in New Zealand.[4]

The Board advertised the chair of Physics, requested Rutherford to be on the selection committee, and Dr. Farr was appointed. (During Rutherford's visit to New Zealand in 1925, in an address to the College, he revealed his attitude towards engineering: "Very few engineers are physicists . . . " he announced. "If you take my advice . . . the more physics you know, the less engineering you need."[5] Judging from the hilarity with which this remark was received, a fair proportion of Rutherford's audience relished such an anti-engineering joke.)

The Board was not successful in inducing Professor Scott to relinquish his "dog-in-the-manger"[6] stand. It was not for nothing that in later years D.B.Macleod, a lecturer in Physics under Professor Farr was to describe Scott as a "mastiff" in comparison with Farr's "poodle."[7] Size and loudness of bark were obviously forces to contend with. A letter from Scott to the chairman of the Board in July 1910 stated: "I have the honour to express my views on a proposal that the teaching of Electricity shall still be continued in the Engineering School but under the control of the newly appointed Professor of Physics." The views were antagonistic. He made much of the overcrowding and accommodation difficulties faced by the School: the lecturer in surveying, W.R.Robinson, having to accept a room on the ground floor of the old Chemistry Laboratory; the lack of Laboratory space for Applied Mechanics; no room in which students could keep abreast of the necessary reading and drawing requirements. He had no hesitation in reminding the Board that "the buildings are undoubtedly the property of the Engineering School which should therefore have first call upon them." (This is one of the situations where Scott was able to make the most of the separate endowment and financial arrangements for the School.) All these arguments were presented as cogently and rationally as ever. Scott's real position appeared in the final paragraph with a considerable soupçon of emotional blackmail:

> Having been for upwards of 20 years in sole charge under the Board of the School of Engineering with results which are known and I think I may say appreciated, even outside New Zealand any proposal embodying joint control is naturally distasteful to me — in the present case it is especially so — I am well acquainted with the methods of the newly appointed Professor and from past experience am certain that his presence in the Engineering School with an independent control would be an impossible arrangement . . . I would ask the Board to pause before deciding on a course of action which must greatly reduce

> the usefulness of a National Institution and destroy the pride and deep personal interest in the efficiency of the School which has hitherto been felt by myself.[8]

(At this stage in the School's history, the phrase "National Institution" reflects only Scott's ambition. The official Government recognition that the School deserved such a title was some years in the future.) Scott was extremely loath to admit an equal authority into his dominion. Indeed, it was most probably this very reluctance which led to his ceding to Auckland the School of Architecture.* Scott retired before it became necessary for him to acknowledge the primacy of the civil engineering section of his School. In 1920, he appointed as lecturer in civil engineering R.A.Campbell, a man of considerable charm and presence. If Campbell had pushed for a chair — although his real interests lay in consultancy not in academia — the numbers of graduates in civil engineering compared with those from other departments would have provided an argument difficult to counter. The more we understand Scott's personality and the qualities he valued in his subordinates, the more we can appreciate how it was that the School was becalmed in the doldrums after his retirement. His successors, Powell and Steele, had been trained in all those years to obey orders, to accept his control and supervision; the very qualities which made them so eminently suitable as colleagues of a difficult superior would militate against their assuming authority themselves with any verve or inspiration.

Scott may have lost the battle over the Physics chair but in other respects he was firmly at the helm. (If any excuse were needed for describing his activities in nautical terms, it would be provided by his own tendency to liberally sprinkle his public speeches with seafaring imagery. According to the *Press*, he gave a "breezy address" on Diploma Day, 13 June 1908, at the time of Professor Cook's retirement. "As senior officer of the ship of Canterbury College . . . he was in duty bound to look back over her course for the two decades during which he had been on board.")[9] Concerned that the numbers of evening students were dropping, Scott went the rounds of the main engineering firms whose apprentices normally attended the School and discovered that the young men were finding the fees too great a burden. A recommendation from him that there should be a reduction of 2/6 for every class after the first two was approved by the Board. In 1907,

*see p. 148

the Christchurch Technical College was established under the auspices of the Department of Education and Scott was very quick to detect any possible encroachment on his territory. There was a certain overlapping of the courses in mechanical engineering, which he regarded as a "breach of faith,"[10] and the Registrar was kept busy over the next months handling the correspondence between the School, the Technical College and the Department of Education.

As the size of the School and the range of courses it offered increased, Scott had to endure the rapid turn-over of staff at demonstrator level, which was to bedevil the School for so many years. Cull resigned in 1908, giving as his reason that he wished to return to outdoor work. Another graduate, R.G.McKay, who was offered the demonstratorship, turned it down because he thought the work load, virtually a 12 hour day, was excessive. Scott rebutted such a notion strenuously, providing the Board with a detailed description of the hours worked by his various staff members. Steele, who had only just taken up a position at the Wanganui Technical School, was sent a telegram urging him to return to the School. (After his graduation as a B.Sc.(Eng) in 1905, he had been a junior demonstrator at the School.) Although he could command twice the salary at Wanganui, he returned at Scott's prompting and thereafter served the School as demonstrator, lecturer-demonstrator, lecturer and finally professor until his retirement in 1946. In effect, he spent almost all his working life at the School of Engineering. Not surprisingly Morrison viewed with alarm the appointment of an engineering graduate to the position of demonstrator, accurately foreseeing that he himself would be passed over in any re-apportioning of salary or status. In 1909, he tried again to secure a salary increase but once again it was blocked by Scott who stated that "neither the character of the work nor the manner in which it has been performed calls for special recognition." In Scott's view there was a clear-cut distinction between the position of demonstrator and that of lecturer: " . . . the term Lecturer [applies] to one who does original work, and the term Demonstrator to one who gives instruction in the work of others." Scott made it quite clear that, in his opinion, Morrison did not deserve to be promoted; indeed, it was "Mr Steel (sic) who [had] some claims to the senior title."[11] Poor Morrison was doomed to remain in a very menial position. His attempts from time to time to gain positions outside the School were unsuccessful, and even these failures could be held against him, a sign that he just was not good enough.

The majority of appointments during these years were graduates of the School. Robinson, the lecturer in surveying, was a notable exception to this rule. He was a first-rate appointment and Scott even accorded him an "Excellent" in one of the confidential staff reports he submitted annually to the Board, the only staff member to be so described. The Board was very quick to see the advantage of having a tame surveyor on its payroll and over the years Robinson and his students surveyed numerous areas of College land. When the land in question was the College block, or even, later, the new site at Riccarton for the Boys' High School, the Board's requests were not unreasonable. But, from time to time, the Board required that some of the endowment land be surveyed, even as far away as Lake Coleridge. Robinson was a very humble man, but at last he was brought to protest. Scott backed him up, suggesting that the Board employ a licensed surveyor as it was not always feasible to include the Board's requests within a teaching programme. In a sense, however, it was Scott's own fault that the Board regarded its engineering staff as unpaid dogsbodies. Scott's enthusiasm for his profession, his keen eye for detail, combined with a great reluctance to see something done badly when he could do it better, led him to offer his professional expertise gratis to the Board on a number of occasions. At a time when he was commanding a salary of upwards of £200 more than any other professor, and this despite the School of Engineering's "depressed state of finance,"[12] he evidently saw no reason to expect a greater return for his services than a letter of gratitude. Later, he was to feel less generous.

CHAPTER EIGHT

War, 1914-1918

It was as well that the School of Engineering completed the extensions for its hydraulics laboratory when it did. A year later, war had broken out in Europe and expansion ceased. Patriotic fervour was very strong in New Zealand; there was considerable loyalty for England and the Empire. The Board of Governors expressed great pride in the numbers of young men, students and staff, who promptly enlisted for active service. The School of Engineering sent many of its young men to the front. As Scott declared in his contribution to the Chairman's 1915 annual report: "Engineering students are, on the average, older than those taking other courses at the College, and from the nature of their training are of very considerable value in the field. It is gratifying to find that the senior students have fully realised their duty and have gone almost in a body to the front . . . "[1] By 1917 "the total number of those recently at the School, and now with the fighting forces [had reached] one hundred and seventy six." Constant regret was expressed that there had been no attempt to "utilise the services of these highly trained men in the Engineers, Artillery or other special corps."[2] The death toll of New Zealanders in active service was very high. Proportionate to the size of its population, New Zealand lost more lives than a country like Belgium which was in the front line of the war.

The war did not greatly exacerbate the difficulties of retaining staff. Demonstrators, as ever, came and went; the lure was not always active service but the higher salaries obtainable from the Public Works Department. T.Dalmer, for example, resigned in 1914 "in connection with [the hydro-electric development at] Lake Coleridge."[3] Another graduate and a very recent appointment, G. MacIndoe, did enlist and went off in 1916 with the good wishes of the School and the Board of Governors. He was killed on active service. Twice, Powell's name came up in the ballot. The first time he was not certified medically fit. The second time, Scott, setting aside a previous Board resolution that there should be no appeals, convinced the Board that, as it would be quite impossible to replace him, the teaching of Electrical Engineering would suffer. The appeal was successful and Powell remained with the School. Scott himself served as Chairman of a Government Munitions Committee throughout the war, with Percy Bamford, yet another graduate of the School, who was persuaded to fill MacIndoe's vacant position in 1916, as his assistant. Special pleading from the Board was once more necessary to retain Bamford's services.

Otherwise the work of the School continued. It obviously became impossible after a while to import any major new pieces of equipment. At first, war risk insurance had to be arranged; later, firms in the United Kingdom were apologising for delays "because of the dislocation of shipping."[4] Plans for new equipment were therefore held over until the cessation of hostilities. But money was in short supply. A letter from the Secretary of Education stressed the need to retrench. Despite the willingness of the Board, and especially of Scott, to co-operate in the production of munitions, the Government decided that New Zealand could best "serve the Empire"[5] by concentrating on the production of its primary products. For several years after the war ended, there were difficulties about importing books and equipment; at first it was not possible to engage in trade with Germany or Austria, although Scott looked forward to the prospect of a return to normality because the exchange rate between New Zealand and the States was prohibitive, whereas that between Germany and New Zealand was very favourable.

Once the war was over, there was a great influx of students, with the attendant problems of severe overcrowding. The Public Works Department intimated that it would resume sending cadets to the School, and every effort was made to assist returned servicemen to complete their interrupted education and resume normal life. "There is every indication that the limit of our present capacity will be reached in 1920,"[6] Scott wrote in the 1919 annual report. A decision to send a deputation to Wellington to submit "the urgent requirements of the School of Engineering"[7] was postponed "on account of the unsettled state of the Government."[8] By this time, Scott had a permanent seat on the Museum and Library Committee; the previous year he had requested a "courtesy seat" in the light of the importance of the School and his long-term standing in the College. (The next year, despite the fact that the School's funds could not stretch to the employment of a junior clerk at £200 a year, so that it had to make do with a typist at half that salary, Scott's salary had been increased to £1,000.) On this committee, he was well placed to direct the Board's negotiations with the Department of Education. First, however, it was necessary to convince Sir Francis Bell, the Minister of Education, that the old Boys' High School building was totally unsuitable for the School's requirements. (Boys' High, having been re-sited at Riccarton, was about to become available for College needs.) Once this suggestion had been scotched, negotiations went ahead smoothly. By July 1920, the Government had granted the School the desired £25,000, £15,000 for buildings and £10,000 for equipment, as well as an increased annual grant of £3,750.

While it would have been impossible for the School to have expanded over these years without Government assistance, the Board had been forced to surrender a great deal of its autonomy in accepting that its School of Engineering had national not just provincial status. (A special committee of the University of New Zealand had been set up in May 1919, to consider, "inter alia, the question of the nationalisation of the Medical School in Dunedin and the Canterbury College School of Engineering."[9] The official recognition of the School's national status was announced in 1920 and Scott could declare that "the School of Engineering [was] now recognised by the Government as the national school of Engineering for the Dominion, a position that it [had], in fact, held for many years."[10]) Whereas previously, the Board had been an autonomous institution by virtue of the endowments established by the provincial government of Canterbury before it was disbanded, it became less independent when a large proportion of its finance originated from the central government. If, earlier, the Board had been entitled to allocate certain moneys for Professor Scott, and to expect from him a detailed and itemised account of his expenditure, now the Department of Education could make the same demands of the Board. Scott, not surprisingly, had chafed and fretted at any restrictions of his power. Keeping and balancing of accounts were a major part of his role as professor in charge. His success in arranging trade discounts on apparatus always made him respond in a very self-righteous manner to any criticism that he had overspent. A charge of exceeding the £900 he had been granted to purchase apparatus on his trip to Great Britain in 1904 was dismissed by him as necessary packing, freight and insurance. Two weeks later he gave a justification of how he had overspent to the tune of

£211.17.0. It was unanswerable:

> I would state that on my recommendations . . . the expenditure of the year having been revised by the Museum and Finance Committees and passed by the Board, such passing and the handing to me of the approved estimates had been considered as authority for my proceeding as required with the expenditure so provided for.[11]

Some restrictions on his assumption of authority were necessary, however. By 24 November 1905, Scott was directed that "any expenditure over £20 be referred to the Finance and Estates Committee and that this refers also to the original vote."[12] When the £20 was later reduced to a mere £10, the cut-off point for the other College departments, Scott found this totally unacceptable. He sought an interview with the Board so that he could present his view that the School of Engineering was different from the rest of the College. Once again he won his point. It was conceded that in the case of the School of Engineering, authority was needed only for amounts over £20.

The relationship between Scott and the Board of Governors in the early days of the School's history foreshadows the later relationship between the Board and the Department of Education. Thus, in exchange for the large sums of money the Government had invested in the School, the Board had to submit to the fact that all decisions involving finance were subject to intense, and not always well-informed scrutiny. Certain aspects of the new building which was to be opened by the Minister of Education, Sir James Parr, in 1923, had been carefully specified by Scott from his knowledge of engineering needs, questions of access, width of corridors and so on. The Department's architect looked over these plans and queried a number of details. In the end, Scott was forced to conclude that the Government architect, who consistently referred to the "pupils" of the School of Engineering, had mistakenly assumed that he was dealing with a secondary school. The modifications suggested by the architect all ignored the actual requirements of a building suitable for teaching engineering to adult students. Then there was the £10,000 equipment grant which was naturally expected to provide for the School's needs over a number of years. Prior to that grant, the Board had been able to apply for a £ for £ subsidy on materials and apparatus, although at one stage Scott had overstretched himself in the particulars he was claiming, and was informed that basic day-to-day items like "matches, tape, dusters, pins, etc . . . "[13] came under the running expenses of the School. Acting on the assumption that this arrangement would continue and that the £10,000 was intended for large items of equipment for the new building, the Board put in its usual claim for Government money. It was not impressed by the suggestion that such claims were no longer valid, that the money was to be taken out of the £10,000. The Department of Education was not prepared to give an inch. If the Board was not willing to accept the conditions under which the money was offered, well then the Board could always refuse that money. Local papers were sympathetic but powerless.

The Department's overseeing of every penny of the Board's expenditure reached such a pitch that in order for one of the engineering professors — this, after Scott's retirement — to order even a small item of equipment, there had to be a letter to Wellington to gain the necessary authority. The whole thing became increasingly time-consuming. If Steele, for example, required permission to spend £34, he had to write to the Registrar announcing this fact. The Registrar wrote to the Department, then back to Steele indicating that this had been done. A further letter to Steele was necessary once the Department's permission had been received. Scott had thought he had been labouring under difficulties when he had to seek permission to spend more than £20 on any one purchase; he would have been exasperated at this further instance of bureaucratic obfuscation.

The "engineering potentate" as seen by Archibald F. Nicoll. The portrait hangs in the School of Engineering.

CHAPTER NINE

Scott's Retirement

By 1921, Scott had turned 60, and his years of frenzied activity had taken their toll. Much as he would have wanted to be Director of the School — a title he had claimed in 1919 — on the occasion of the opening of the new wing, described as "a noble monument to the vision and force of its founder,"[1] his health was not equal to his remaining in charge for a further 18 months. (Indeed, he was not able to attend the occasion, which coincided with the 1923 Jubilee Day celebrations, even in his capacity as grand old man, as he was at the time a patient in the Rhodes Convalescent Home on Dyers Pass Road.) Accordingly, on the grounds of "increasing physical infirmities,"[2] he asked for, and was granted, a year's leave of absence on full pay for 1922, retiring formally at the end of that year.

Understandably there was a great carry-on. The members of the School of Engineering, at a special ceremony, presented him with a gramophone. Nautical to the last, Scott regretted his separation from the School but admitted it had to be, claiming that while "the old hull had been badly battered in its thirty-six years' voyage . . . it had reached a safe anchorage."[3] At a special dinner given in his honour by the Board of Governors, Scott was again surrounded by accolades and much speechifying. In return he attributed his success to the training he had received in his early life, expressing appreciation that this had occurred "before the days of specialisation."[4] Next, in May 1923, the Board appointed him Professor Emeritus in recognition of his eminent services over the years.

By granting Scott a year's leave of absence, the Board gave itself a year in which to decide the best course of action as far as the re-organisation of the School was concerned. As he was still employed by the Board, he accepted responsibility for ensuring the smoothest possible arrangements for the School for 1922. The Board's letter to Powell, who on Scott's recommendation was made Acting-Director for the year, followed almost verbatim Scott's own letter to the Board. He was able to recommend a most parsimonious re-organisation, shuffling various staff members like Bamford and Steele around, but the main money-saving device was that he passed on to the staff his lectures complete with diagrams and sketches, while at the same time carefully retaining copyright. In his opinion his lectures on "The Theory of Workshop Practice" constituted "an almost unique course."[5] From his correspondence with the Board at this time, it becomes clear that almost all the lectures given at the School had been written by Scott and were merely delivered by his underlings, which did not augur well for the future of the School. This will account for Scott's anxiety about the copyright and safety of his lectures when he had to hand them over for others to deliver during his absence. 250 lectures covering seven different courses were listed in 1904, with anxious instructions to make sure none of them went astray. An anecdote which went the rounds of the School suggests the possibility that some, at least, of these lectures survived even after Scott's retirement. One particularly heavy afternoon, Percy Bamford, described by an ex-student as "a dead loss"[6] as a lecturer, was droningly reading a lecture and, half asleep, neglected to observe that the sentence he was reading contained an instruction to the lecturer, not to the students. "Here, tell a funny story," he announced and carried on with his reading, much to the students' amusement. (This story, in various versions, became a familiar chestnut of the School over the years.) As there is no evidence that these lectures were up-dated from year to year, we may well question the quality of the teaching — L.B. Wood certainly left it on record that Scott "was not a brilliant lecturer."[7] The *Press,* describing one of Scott's "Popular Lectures" on the subject of Battleships, claimed that it was "bristling with technicalities"[8] but intelligible to the lay person. A certain amount of Scott's effectiveness would necessarily have been in his delivery and appearance, an "Engineering potentate, bulging at the waist-coat,"[9]*

*J.H.E. Schroder was the author of the verse, a parody of Masefield:
"Engineering potentate, bulging in the waist-coat,
  Clumping down the corridors with firm, proud tread,
  With a glare for his lecturers,
  Morrison, Bamford,
Filling all his greasers with stark, chill dread."

according to a student song, as well as in his enthusiasm for his subject. But lecturing to undergraduates is very different from addressing a public audience.

Anyway, for the year 1922, the School of Engineering coped, with Powell and the rest of the staff endeavouring, as far as they were able, to preserve the status quo. This meant that the Board could give its attention to deciding what form the direction of the School should take without Scott to lead it.

At first, the most favoured plan was to advertise in the United Kingdom and Australasia for a Professor of Mechanical Engineering, who should also be Director of the School for a three-year term at a salary of £1,200 a year; in other words, Scott's successor was to step exactly into his shoes. Inevitably there were risks attendant on such an appointment; in Scott the Board had had not only a man with extensive design experience and enthusiasm for his subject, but one very much respected by the engineering profession in New Zealand, so many of whom were his past students. As well, he was a man with superb P.R. ability, whom L.B.Wood described as a "most persuasive 'beggar'."[10] Even a committee set up to consider Auckland's claims for a School of Engineering was forced to concede that " . . . much valuable apparatus was given to the School [over the years] either free of cost or at greatly reduced figures, owing to the genius of . . . Professor R.J.Scott in obtaining the sympathetic interest of manufacturers in England in the work he was doing."[11] The chances of procuring a man equal to Scott were very slight, and, inasmuch as there did not appear to be anyone in New Zealand whom he could recommend, the Board faced the additional disadvantage that the new man would be an outsider, who would naturally take some time to adjust to New Zealand ways.

Inevitably Scott was consulted about the future of the School and gave as his opinion that "the ideal arrangement for a National Institution would be a completely self-contained School under a Professional Head, teaching *all* subjects required by a Student after passing the Entrance Examination and granting a Certificate on the satisfactory completion of the Course." He fully realised that it was not feasible, but he would have liked "the University Degree in Engineering to be granted only to those candidates holding a pass Degree in Arts or Science and the Certificate of the Engineering School." However, one of the problems about deciding on an appropriate re-organisation of the School was the ambiguity of its relationship with the College:

> Had I been able to continue in charge of the School the present somewhat indefinable position with reference to the College Proper might have been maintained for some years by a policy of mutual give and take, but always with the danger of friction arising from the dual control of the Student . . . .
> With the appointment of a new Director, especially if a Professional man not used to the academic outlook a difficult position would be created; and since complete separation cannot be obtained, fusion appears to me to be the next best course . . . .

Pragmatic as ever, Scott, therefore, recommended that four chairs be established — Scott included Robinson, obviously recognising the importance of surveying within the engineering world—and that Powell should be Dean and Head of the Engineering School as "the best arrangement under the circumstances."[12]

During the time of the Board's deliberations, when it still seemed likely that an outside man would be brought in as Director, in which case the existing staff would continue in their present menial positions, Steele and Campbell, in a letter endorsed by Powell, expressed their uneasiness. "The advancement of engineering knowledge during recent years has been so rapid that nowadays, anyone claiming to be an expert in more·than one branch of the subject would have his bona fides severely criticised."[13] Scott could boast that he had been trained "before the days of specialisation"[14] but his era was over. While it was to be years before post-graduate research in engineering would become a *sine qua non* for the appointment to a University lectureship, this was the direction in which engineering education was headed, towards a more specialised degree. Scott himself did not have a degree at all but he was professional to the core; Arthur Dudley Dobson described him as "a true engineer, who lived up to the highest ideals of the profession."[15] Having made their position clear — they feared "redundancy of the teaching of one subject" — the three lecturers pointed out, with justice, that the Board's policy in debarring them from taking outside work meant that their prospects of finding advancement outside the School were very slender, as indeed has been evident from Morrison's abortive attempts to procure employment. Thus "if they [were] to be only able to look forward to holding an inferior position, it [would] be hard for them to maintain their interest."[16]

It is always a matter of regret that so many of the discussions behind

Board decisions should be lost when only resolutions are recorded. We are indebted to L.B.Wood, who from time to time felt so strongly about a particular issue that he circulated his views in writing before a crucial Board meeting. This was the case before the meeting which resolved to promote the three lecturers and appoint a Dean rather than advertise for a Director. Wood felt the whole exercise was very dubious; he questioned the advisability of promoting the three men — the possibility of offering Robinson a chair was never considered — without advertising and providing free competition, and cited the recent case of Professor Condliffe, when the "Englishmen on the Professorial Board [had] successfully insisted that the chair should be advertised and the candidates scrutinized by a committee in England."[17] As well, he seriously questioned the suitability of these men for the new positions, mentioning that on the whole all that Scott could say in their favour was that they were good teachers, and even that he doubted when they were, in fact, merely reading Scott's lectures. (Wood's zeal was getting the better of him; this was an exaggeration. Both Powell and Campbell, originally appointed as lecturers, would have been responsible for writing some at least of their own lectures.) Scott's admission that they were good lecturers, with an added, but tepid, recommendation for Powell, was given verbally at one of the sub-committee meetings convened to discuss the re-organisation of the School. According to Wood, Powell's "administrative ability [was] practically an unknown quantity and his personality [did] not lead one to believe that he would excel in persuasiveness."[18] A voice crying in the wilderness, Wood was unable to change the Board's decision to rescind the original motion about advertising for a director, and the triumvirate, Powell, Steele and Campbell were elevated to chairs as from 1 January 1923.

By his premature retirement, Scott certainly placed the Board in an invidious position. Maybe, if it had had several years' warning in which to consider what was the most appropriate course of action, it might have been possible to appoint someone who would inject new life into the School, someone strong enough to make a complete break with the Scott tradition. With hindsight, it can be seen that the promotion of the three men to chairs was an unwise move which ushered in a period of increasing inertia. The School was becalmed; the challenges which the next twenty-eight years were to offer — the depression, the second world war, and the drastic overcrowding which followed it — came mainly from outside the School. Instead of the great peaks of activity which characterised Scott's reign, the School now had to struggle against the limitations of circumstance. It was to continue in this way until the move to Ilam.

The interim period when the Board was deliberating was a year of limbo, with Scott hovering rather like Banquo's ghost in the wings. He claimed, for example, that as he had not yet officially retired, he was still "morally and legally"[19] entitled to receive the half share of the School's testing fees for 1922, £39.18.6. Seen in isolation and no longer in the context of a flourishing School of Engineering, this insistence on his due sounds very shrill. He was paid the £39.18.6. Indeed, Scott's obsessional belief that he had been financially deprived considerably soured his relationship with the Board in the first years of his retirement. It is the case, however, that his decision to retire would not have been lightly taken. As the amount of superannuation was calculated on a proportion of the salary of the last three years, by retiring early he would have to accept a reduced rate. To minimise the inconvenience of this, he requested the year's leave of absence on full pay, although request is rather a euphemistic term for what was a piece of emotional blackmail. Scott guaranteed to retire at the end of 1922, provided the Board acceded to his request. Otherwise he would carry on with the imminent risk of a nervous breakdown, to the detriment of himself and the School. The Board, as we have seen, complied. Unfortunately for Scott at this point the Government intervened, passing a Public Expenditure Adjustment Act by which his salary for 1922 was reduced by £103.10.0. His sense of ill-usage was immense; he even went so far as to accuse the Board of going back on its word. Letters went to and fro between the Board and the Department of Education; there was no way the Government was going to modify its position. If the Board wished to grant increases of salary to deserving staff, after first having reduced them according to the Act, that was its affair. In the event, Scott's salary was reduced only by £50, which was a considerable concession, but still he was not satisfied.

Firmly believing that the Board still owed him £50, Scott submitted an additional itemised account for £186 which he claimed was due to him for his professional services in connection with the new College heating and lighting scheme. It had been recognised for years that students and staff were being asked to endure arctic conditions in the lecture rooms and laboratories. The *College Review* in 1904 had reported:

> An interesting experiment during an electricity lecture was recently brought to a standstill owing to the sudden freezing of the water in the battery. During another lecture lately the mercury stood poised at the giddy height of three degrees above freezing point. The feeling is growing among students that a little more heat would make Rooms A and B of the Engineering School more comfortable.[20]

However, there was considerable delay before the Board accepted Scott's offer to submit specifications for a new heating system and even longer delays before these were implemented. It was not until 20 years later that the new system was installed under the supervision of Professor Steele, designated in a letter to the contractors as Engineer to the Board. The delay had been caused by the difficulties of importing the necessary equipment during the war and not, as in later years, by bureaucratic dilatoriness. At the time when Scott had presented to the Board his specifications — which, in fact, had been prepared by his students — he had gratefully received its letter of appreciation with seemingly no thought of payment in his head. Now he was submitting a claim. The Board, obviously uncomfortable, presented Scott's itemised account to the College lawyer, F.I.Cowlishaw, who carefully checked all the relevant letters, even the terms of Scott's original appointment in 1889, and gave as his opinion that the Board was not legally liable. While it would have been possible under the schedule of appointment for Scott to have refused to undertake work of this nature, it would have been quite out of character. By his enthusiastic response to such technical matters as the erection of the College flagpole, Scott had accustomed the Board to expecting that its engineering staff would be regularly on call for professional services and advice. In this way a precedent had been created which later staff of the School must surely have resented. In this instance of the new heating scheme, it was eventually agreed that Steele should be paid a percentage of the cost of installing the heating system and in all fairness, as Steele's plans had been computed on Scott's calculations, Scott was offered £60 — not the £186 he had claimed. The whole episode is a sorry conclusion to a splendid career.

But that it was a splendid career was never in doubt. It was said of C.E.Bevan-Brown when he retired from the headmastership of Boys' High in 1920, that he had "become an institution, the personification of the school."[21] Scott was deserving of a similar eulogy. From the most humble and tentative beginnings, when it served the needs of a few apprentices, the School had grown in reputation and prestige, offering degrees in the three branches of engineering, recognised by the Institution of Civil Engineers, and providing qualified men for engineering trades and industry. In the thirty-six years since Edward Dobson gave his inaugural address, over 2,000 students had passed through the School. "It was fitting that the culmination of his work should be the recognition by the Government of the School as the National School for the Dominion."[22]

Despite the precarious state of his health at the time of his retirement, Scott lived on until 8 November 1930 and was still fit enough to be sailing his yacht *Boojum* the year before his death. He was accorded an impressive funeral, with a highly dramatic flavour.

> From 10 o'clock till 11 the body of Professor Scott lay in state in the College hall. As the College clock chimed 11 the two minutes' silence to commemorate Armistice Day was observed. The flag was run down to half-mast and hoisted again when a blast from the bugles at the Cathedral gave the sign for the termination of the silence. The cortege then proceeded by way of Park Terrace and North Park road to the cemetery.[23]

Tributes to Scott from previous students poured in. "One and all will mourn the loss of a great master, and the memory of his charm of personality will be treasured for long."[24] When F.T.M.Kissell canvassed ex-students of the School the world over in 1934 for subscriptions towards a portrait of Scott, he was most gratified at the response. A.F.Nicoll painted the portrait from a photogragh and Kissell presented it to the School. It still hangs in the first floor of the administrative block of the School at Ilam. The *Press*, summing up his years of service to the School, declared that Scott's work "began with its foundation, was continued for more than thirty years, and established its own praise in the reputation of the School."[25]

CHAPTER TEN

Organisation of the School: A Director or a Faculty

Scott's retirement in 1923 left a vacuum. His forceful personality, his strong need to present himself with challenges and then ride them out as courageously as he had once brought his yacht safely over the Dunedin mole, had produced an atmosphere which was exciting and innovative. Sadly, those who came after him were not able to sustain such a creative impetus. Reliable, competent but unimaginative would be the most telling adjectives to describe Professors Powell and Steele, who, between them, "directed" the School during the next years, alternating in the role of Dean of the Faculty.

The creation of a Faculty of Engineering brought the organisation of the School more into line with the other departments of the College. Its affairs were no longer handled by the Museum and Library Committee but, instead, by the College Committee. In this way the special status enjoyed by the School was considerably lessened. Scott had requested a courtesy seat on the Museum and Library Committee. Once this had been granted, it became automatic that when there were engineering matters before that Committee, Scott would attend the meetings. This privilege disappeared and it took several letters from Powell before the Board agreed that the presence of a professional engineer might be a valuable asset to the College Committee.

There was one significant difference between the running of the Engineering Faculty and the other College Faculties. A Professorial Board recommendation in 1924 was accepted whereby the members of the Engineering Faculty had the right "to nominate annually to its membership prominent members of the Engineering profession."[1] In this way, the close relationship between the School and practising engineers was formally asserted. Those nominated in 1925 were: F.W.Furkert, Engineer in Chief, P.W.D., Wellington; A.R.Galbraith, City Engineer,

Christchurch; E.Hitchcock, Electrical Engineer, City Council, Christchurch; F.T.M.Kissel, Chief Electrical Engineer, P.W.D., Wellington; and G.S.Lynde, Chief Mechanical Engineer, N.Z.Public Railways, Wellington. The following year, Cyrus Williams, who had assisted the Civil Engineering Department for many years by giving lectures on Harbour Engineering, was added to the list. (The presence of non-academic engineers at the faculty meetings conferred a benefit which was not confined only to the School. E.McKillop, the retiring Commissioner of Works, maintained in 1955: " . . . because of the dependence this department must place on the National School of Engineering, it is very desirable that the existing relationship [between 'the engineering administration' of his department and the faculty] should continue and, if possible, be strengthened.")[2]

Several of these men had been consulted in 1923 and asked to give "criticisms or suggestions" concerning the course for the B.E.(Electrical). "The Board of Governors, being anxious that the School of Engineering should keep abreast of advances in the science and practice of the profession has decided to seek the advice from time to time of professional members of the different branches as to the suitability and efficiency of the prescriptions laid down by the University."[3] While expressing great satisfaction at what the School had achieved, most of the reports concurred that it was most essential that the teaching of the School should be slanted towards New Zealand needs, not towards work "in the older countries . . . Every industry and engineering work of any magnitude ['in the Dominion'] has a . . . range of problems which can only be solved by the academic mind and in which Canterbury College should take its part if it is to build up the reputation we expect of it," wrote Lawrence Birks, Chief Electrical Engineer of the Public Works Department who had served on the staff of the School very briefly in 1904. He also noted the absence of research papers on engineering subjects, a very serious "blank . . . to the status of the engineering school of Canterbury College."[4] It was to be many years before research became a top priority of the School. In his submission to the Government-appointed Reichel-Tate enquiry into University education in 1925, Steele freely acknowledged that pressure of routine work totally inhibited his undertaking any research work. The enthusiasm with which Scott had accepted the two-fold nature of the School, as well as his willingness to undertake complicated testing procedures for industry or Government, in a sense backfired during these years. And once the depression exerted its stranglehold, stifling expansion and creativity, there was to be even less opportunity for initiating original research.

The new management of the School may have been competent but it was dull; however, there was nothing the Board could do about it. Then it happened that in 1929 Campbell resigned his chair to return to private practice, claiming that he had never been much of a teacher, and "didn't look the part".[5] For all that, student memories of him show a colourful character obviously willing to regale them with yarns about his war experiences. The vacant chair of Civil Engineering gave the maverick L.B.Wood, his chance to remind the Board that without Scott to direct it, the School was floundering. Unfortunately an inattentive reporter from the *Star* gave out that it was the Department of Civil Engineering which was inefficient and poorly directed, a criticism of Campbell. This error created a brief furore and necessitated an apology to Campbell for the allegations of inefficiency.

It is easy to see how the word "civil" should mistakenly have been included in his report. To the lay public, if engineering was not military, it was indisputably civil; civil, therefore, embraced all three branches of engineering instead of referring to its own specialism. Even within the profession there was still this confusion in that, until 1937, the national body of engineers was called the New Zealand Society of Civil Engineers, despite representations that such a title excluded engineers of the other branches. At the 1929 annual meeting of the society, there was a referendum to assess whether members wanted the word "civil" to be removed from the title. Although the decision was negative, one of the delegates recorded that despite the fact that he had voted the previous year to retain it, he now felt that the dignity of the society would not be endangered by the omission of the word "civil". It is not surprising under these circumstances that a casual reporter should get it wrong. Unhesitatingly, the Hon. G.J.Smith, Chairman of the Board, wrote to the *Star* rebutting Wood's criticism. He asserted that "the three professors have done excellent work and the reputation of the School of Engineering for teaching and results still stands very high."[6] Smith indicated that the Board, after a full discussion, had rejected Wood's suggestion that it take the opportunity of Campbell's retirement to appoint a Director, "a better man than the ordinary run of professor."[7] As he explained, "Professor Scott had an unique personality . . . But to get a man of similar personality and attainments would be very difficult and certainly he could not be got at double the salary the college can afford to pay."[8] Wood in "the course of a chat" with a representative of the *Press*" accused the Board of "boggling"[9] at the appointment of a controlling head. But if he was willing to ignore the economic pressures of the times, the Board, responsible for the major educational establishments of Christchurch could not lightly sign away the extra money which would be needed to attract a suitable applicant. Already, there was considerable evidence of hardship in the city. When the advertisement appeared for the position of caretaker at Avonside Girls' High School, formally established in 1927 under the Board's auspices, the applications had revealed " some pathetic cases of unemployment . . . "[10]

Needless to say, the Board was motivated not only by a sense of economic responsibility in deciding that it would not advertise for a Professor of Civil Engineering who would at the same time be the Director of the School. There were other factors to be considered as well. At the slightest suspicion that they might have to endure another Scott in their midst and return to the positions of underlings in the School, Powell and Steele, supported by Campbell, once again wrote protesting that "it is absolutely impossible for one individual to be thoroughly conversant with all branches of Engineering." They also reminded the Board that the appointment of a Director would mean that "there would be no need of a Faculty of Engineering".[11] This would inevitably create a most ambiguous relationship between the Director and the other Professors. But it was most likely a memorandum from Dr James Hight,* the Rector of the College, which clinched matters. He strongly queried the £1,000 salary. Were the responsibilities of the Civil Engineering Chair more onerous than those of other Chairs in the College that it was deserving of the higher salary? Or was the Board tacitly recognising that the standard salaries of Chairs were too low? "The decision to make the salary £1,000 will inevitably re-open the general question of salaries which the Board may be called upon to face"; a statement that the Board would have found most unpalatable. As well, Hight fully recognised the necessity of safeguarding the interests of the existing staff. He recommended that if the dual position were to be advertised "the following considerations should be taken into

*Hight, James, 1870-1958; B Halswell; educ Malvern District High School, CC, (BA), 1893, (MA Ist Cls hons in Eng and Fr), 1894; lect const hist and pol econ, 1901-5; lect hist and econ, 1906-8; dir of studies commerce, 1906-20; LittD (first in NZ), 1906; prof hist and econ, 1909-20; prof hist and pol sci, 1920-49; second rector of CC, 1928-41; m BG, 1923-42; m Sen (1912-48) and Pro-Chan (1935-48), Univ of NZ; cm Acad Bd, 1934-48; CMG, 1932; KBE, 1947.

account: Does it not imply a want of confidence in the present professorial staff? If so, is this warranted by the facts? Does it not assume that the Chair of Civil Engineering has superior claims to be associated with the directorship of the School to those of the Chairs of Electrical and Mechanical Engineering?"[12] It seems that the Board was given no option but to advertise throughout Australasia for a Professor of Civil Engineering and dispense altogether with the notion that it should appoint a Director.

The problem as to whether one Chair should have more claims than the others was revived in 1941 when the whole question of appointing a Director was once more under discussion. It would appear that the Council's reasoning went something along these lines: "during the Scott era, when he virtually directed the School single-handed, the School had undoubtedly flourished. Now that it has fallen into disarray, what is needed is a strong personality to assume authority." But this was to ignore all the other factors which had led to the School's gradual diminishment, especially the drastic shortage of money to fund any enterprising new development. Wood had described Scott as " 'a most persuasive beggar',"[13] but even his blandishments would have been of no avail against the economic struggles of the depression years. Scott was also fortunate in that mostly he wanted money for large, dramatic and newsworthy ventures — the Electrical Laboratory and then the Hydraulics Laboratory — not for general upkeep and maintenance, which, in the nature of things, excited little interest.

Still, the idea of a Director was mooted again with the appointment of a new professor of Electrical Engineering when Powell retired, a kind of faded action replay. The idea was scotched once and for all in 1947, when R.J.Rastrick, the newly appointed Professor of Mechanical Engineering, commented acerbically that he objected to the suggestion that the new Professor of Electrical Engineering, Powell's replacement, might possibly be appointed Director of the School. Apparently no such option had been offered to him the previous year when he was appointed to replace Steele. Realising that it had placed itself in an invidious position, the Council gave up the thought of discovering a second Scott amongst its applicants. But the idea had died hard.

Part II: In the Doldrums

CHAPTER ONE

At a Standstill

Dissatisfaction at the way the School was developing was not overt in the early years after Scott's retirement; on the contrary, there was still a great deal of pride and complacency. Powell spent 1924 in Great Britain, where he "purchased apparatus" to the value of £6000. Like his predecessor, he visited widely, "ten colleges and thirty-four works, in addition to various museums, Wembley, and the National Physical Laboratory." Questioned by a reporter about the relative positions of Canterbury and its English counterparts, he maintained: "the Canterbury College School of Engineering had not much to learn from the universities in England."[1] Similarly the second report in 1926 from R.A.Lawther, holder of the Engineering Exhibition, which was established in 1919, at the General Electric Company in Birmingham, confirmed this self-satisfied view. "As a result of the little investigation I have carried out," he wrote, "I consider the course as given at Canterbury, is very sound indeed, and I should be proud of it in any Engineering Circle the world over." He claimed that "thoroughness as taught at Canterbury, should stand a man with a necessary degree of raw intelligence to work on, in good stead; thoroughness being so important in engineering practice."[2] Against this, we have the astringent comments of Dr R.H.Grundy, who resigned in 1929, having been refused one year's leave of absence "with or without financial assistance, in order to return to England to take up a refresher course."[3] Grundy had travelled out from England in 1924 at his own expense in order to gain experience but after four years, he was greatly disillusioned:

> I found . . . that this experience consisted of teaching during the evenings to a standard which is largely that of an elementary school, and during the day with equipment which in my own subject

of Heat Engines is to a great extent obsolete and totally inadequate. The salary moreover, compares unfavourably with that of a schoolmaster in a Secondary or Technical School. I therefore felt that to one who has had experience of live modern educational establishments in England, the prospect of continuing here at such a salary without refresher courses was intolerable.[4]

Steele, in commenting regretfully on Grundy's resignation, claimed that the latter had always seen himself as "in exile"[5] here in New Zealand. But there is sufficient evidence from other sources for us to be certain that his jaundiced view was not without cause. As he claimed: "the positions below that of a Professor in the School of Engineering here are, in the main, blind-alley occupations; because the routine work, set out to a written time-table calculated in half hours, with neither extra time nor facilities to keep up to date, together with the lack of contact with outside engineering work, makes one unfit to take up a responsible position elsewhere."[6] After all, he was merely saying what the three incumbent lecturers had said when they thought they might have been passed over when Scott retired, and what Morrison had said on several occasions. Grundy could only conclude that the Engineering School did not pay for qualifications. (His was a rare case; the Ph.D., a *sine qua non* of today's appointments was actually a disadvantage. In the 1920's he was over-qualified for the School of Engineering.)

One way in which we can assess the quality of the teaching offered by the School during these early years after Scott retired is to scrutinise the overseas examiners' reports, which were fairly detailed and specific. Over the years, the examiner for the Mechanical Engineering portion of the course, Professor A.L.Mellanby, indicated that for most of the papers "the general results were quite good and that the candidates showed a high standard of attainment."[7] Mechanical Drawing regularly received high praise. Equally regularly, Mellanby deplored the fact that the students were not confident in using the steam tables. "The proper use of [these] requires to be taught, especially in the matter of estimating heat-drops. Candidates would persist in using a formula for the Rankine cycle instead of reading off the heat quantities from the tables."[8] In 1926, he expressed his disappointment that "in the paper The Steam and other Heat Engines (Advanced), no candidate attempted the question dealing with 'Supersaturated Steam'. In view of the great attention that has been given to this subject in the technical press during recent years it was hoped that this question would give an opportunity of showing some acquaintance with current technical literature."[9] By the time he was reporting on the 1926 examinations, Mellanby was obviously exasperated that there had been no attempt to modify the course or the teaching of this paper in accordance with his recommendations. He had become very outspoken:

> The impression gathered was that instruction must have been given on the lines suggested in old fashioned textbooks and that modern methods of calculation, using the facilities offered by the Callendar steam tables, had been neglected. It should be pointed out that the cumbersome formula which was generally used for Rankine cycle problems is no longer necessary and also that it is not sufficiently accurate for present-day use.[10]

This is a considerable indictment, revealing the dangers of "reading" Scott's lectures without any attempt to upgrade them or keep abreast of new developments.

The Civil Engineering papers were examined by Professor F.C.Lea, who remarked in 1925 that "there was nothing to report of a special character except that the work is not quite up to the high standard of three years ago and this seems to be similar to what is generally felt in regard to work of Engineering students in this country."[11] However, the following year he spoke out more strongly, giving his opinion that "the theoretical work leaves something to be desired."

> There is far too much tendency to commit formulae to memory. Taking for example a problem on a concrete beam the candidates wrote down formulae, evidently committed to memory from an American report which had been drawn up as a guide for office purposes. Students should not be walking Engineering "Pocket Books." No student gave the slightest hint as to how the formulae were derived or defined the terms. A sound knowledge of principles is the true basis of a University course, and I suggest that students should be encouraged to work all problems from first principles. Pocket books and other aids are available for design purposes.[12]

The same need for a sound knowledge of "fundamental principles" and "more discipline in calculations for practical ends"[13] was the tenor of his comments in succeeding years.

The examiner in Electrical Engineering, Professor Miles Walker, wrote especially to the University of New Zealand in 1925 commending the work of "Bayonet" — the papers sent to England were anonymous and each one was given a specific code name taken serially from a dictionary — and suggesting that he apply for one of the £100 Research Scholarships at the College of Technology at Manchester. "Bayonet", which was the code name of J.E.L.Caldwell, was an exceptional student. Walker's other comments revealed the importance of encouraging the student to acquire "the habit of thinking for himself in the applications of electricity." He stressed "the desirability of giving the students more opportunity of experimenting for themselves and putting them on their own resources."[14]

Sifting through the evidence offered in these reports, we can see that the teaching offered in the School was adequate but steadily becoming out-of-date, with little encouragement given to the student to think for himself. However "modern" Scott's original courses might have been, if the current teachers did not upgrade their lectures, then the School's standard would inevitably fall. And indeed, this appears to have been the case. In 1934, Powell quite unselfconsciously acknowledged that the courses were the same as in Scott's day, seeing in this merely a tribute to his master's genius and no reflection on himself or his colleagues. It was particularly alarming that on the whole the greatest indictments were directed against the Mechanical Engineering papers, when these provided the foundation of all three engineering degrees. Although, in order to resist the appointment of a director, the three professors cited the fact that engineering now had distinct specialisms, the B.E. degree at Canterbury in these years still retained its predominantly mechanical flavour.

This bias towards Mechanical Engineering was by no means generally acceptable. In 1922, during the considerable upheaval which followed Scott's premature retirement, Campbell submitted to the Chairman of the Board a memorandum of suggestions for improving the efficiency of his department: "If the University only granted a general degree in Engineering and students were allowed to specialise in certain subjects in their final year, the argument for amendment of the course would not be so strong; but as a degree is granted in *CIVIL* Engineering, the graduate should leave the College with more knowledge of the Principles and Practice of that branch of the profession than it is possible to give him under the present syllabus." Directing attention to the descriptive handbook issued by the School of Engineering, Campbell complained that so many of the courses prescribed for Civil Engineering were set down "as for Mechanical Engineering. That is to say that with the exception of Surveying the Civil Engineering student in his third College year (second professional) is really only taking subjects of the Mechanical Engineering course." Third professional fared no better:

> In the third professional year . . . Hydraulics and Pneumatics is "as for Mechanical Engineering" the details of which . . . do not include any reference to the most important part of the subject from a Civil Engineer's point of view: namely, the determination from rainfall and the natural features of a watershed of the run-off or water available for supply or power.[15]

If we set Campbell's criticisms alongside the strictures issued by Mellanby on aspects of the Mechanical Engineering course, we can begin to understand how the School deteriorated over the following years. Although Campbell protested, and his successors Cull and G.G.Calvert likewise protested, very little was changed. When H.J.Hopkins* arrived in 1951 to take up the Civil Engineering chair, he was amazed at the subjects which civil engineering students were expected to study. There was no one strong enough to battle with the Council** so that it would recognise that many improvements and modifcations should be carried out; and shortly, to do it justice, there was not the money. As well, Steele and Powell had set themselves to maintain Scott's standards but failed to recognise that, in this way, the School would stand still. The theory and practice of engineering throughout the world was expanding, but the Canterbury College School of Engineering was isolated in a backwater of its own.

*Hopkins, Henry James, 1912-86; educ Guildford Grammar School, Univ W. Australia, (BE, BSc), Rhodes Scholar, Brasenose College, Oxford (MA); Chief Structural Engineer Courtaulds, Ltd (Coventry); sen lect Univ of W. Aust, 1948-51; prof CU, 1951-78; prof emeritus. Annual Hopkins Lecture established upon his retirement by CU and IPENZ.

**The Canterbury University College Bill, a consolidation of the 14 Acts relating to the College from 1896 to 1928, was passed at the end of 1933. It changed the name of Canterbury College to Canterbury University and the Board of Governors became a Council.

J.E.L. Cull

S.Steele

R.A. Campbell

P.H.Powell

CHAPTER TWO

The School of Engineering: "A Rabbit Warren"

From the outside, the grey stone buildings of the School of Engineering made an imposing spectacle. Scott and the Board of Governors could justly be proud of them. This "legacy of the past" was to become, however, "a heavy handicap."[1] For as the years rolled on, the numbers of students increased which necessitated more staff and more staff accommodation; plant and equipment became larger and heavier; and the buildings proved most inconvenient and inefficient. According to ribald statements in *Canta*, students from other faculties of the College justified the separatist behaviour which kept them apart from engineering students on the grounds that, having once set foot in the School, they became hopelessly lost, that appeals to engineering staff or students were of no avail as they were unable to find their way round the "rabbit warren"[2] themselves. It was claimed that the arts student "is forced to recognise the presence of his mysterious [engineering student] brother — witness the congregation of pestilential vapours which plunge the quadrangle and its environs into pitchy darkness at noon. But he has no knowledge of these infernal regions, no passport into this unknown realm."[3] To which an engineering student replied: "The Engineering School may be dim and mysterious to other students, but we are afraid we cannot give them the freedom of our corridors as they would get lost. We do ourselves, chiefly 'the morning after'."[4]

But while the students found it a matter of jest, a 1944 report on the shortcomings of the School was less humorous: "the limited space and the faulty lay-out combine to make the organisation of instruction difficult, to multiply lectures and laboratory work, and to create many complications where planned order should facilitate the activities and set an example of engineering efficiency to students and others."[5] Scott would have been mortified, but the truth was that the policy of "haphazard changes"[6] over the years had made for total confusion. There just was not enough room, and so a policy of ingenious makeshift and compromise prevailed. A graduate of the 1940's strongly endorsed the very point of view which Powell had vainly endeavoured to make the Council accept, when he complained:

> To start at the beginning, the approach stairs from the quadrangle on the west side . . . are situated in a poorly lighted area — the stairs themselves are narrow and unimpressive, not at all befitting the dignity of a modern seat of learning. From the head of the stairs the corridor is narrow and bare, with poorly lighted corners, and its extension passes through a change in level to approach the private rooms: all this area is drab in the extreme.[7]

The Council was not totally indifferent to the plight of the School but there was no money available for the major restructuring which was so obviously necessary. And by this time, it was hamstrung by having relinquished power into the hands of the Department of Education in exchange for financial assistance. The institution which had so proudly been heralded as the "finest South of the line"[8] in 1910 was now deserving of the description "ramshackle."[9] The drastic overcrowding caused by the influx of returned servicemen after the Second World War led to a highly controversial debate as to the School's future. By that time Powell was thoroughly exasperated with what he described as the piecemeal development over the years, the patchwork of additions and accretions. In a letter to the Council in 1945, he spelt out the situation quite specifically. His report the previous year had referred to a lack of "planned order;"[10] it would be nearer the truth to talk of a planned disorder:

> A glance at the arrangement of the School shows:-
>
> 1. The apparatus pertaining to the Heat Engines Laboratory is scattered. Part of it, the steam engine, is in the strength of materials laboratory. The gas and oil engines are in another room, the refrigerator and air compressor are in another room and auxiliary gear in another room.
>
> 2. The Hydraulics Laboratory is in two parts at opposite ends of the building.

3. Electronics* is separated from the Electrical Engineering Laboratory.[11]

By the same token, in the years immediately preceding the move to Ilam, staff accommodation became impossible; many staff members had no private room for lecture preparation or for consultation with students, and those who did were often making do with a tiny space partitioned off in a corner of the Electrical Laboratory or the Library. In 1946 a scheme was devised which would allow several staff rooms to be created: partitions were to be erected along one of the corridors so that various junior staff could have separate accommodation. Powell recommended:

> The most obvious and suitable place to put further staff rooms would seem to be the long corridor in the wing joining the old Boys' High School. The corridor is 10 ft. wide and 14 ft. 9 in. high with an available length of 45 ft. If the present cupboards were removed a series of rooms could be built along the inside wall, leaving a 3 ft. passage against the present windows. The rooms could be built with walls of fibre board and glass and fibre board ceilings . . . The rooms would be 6 ft. 9 in. wide and 12 ft. (if there were 3) or 10 ft. 6 in. (if there were 4) long. The room height could be about 8 ft. 6 ins.

Powell justified the creation of these horseboxes by citing the proportions of some of the other staff rooms, in comparison with which "the suggested width of 6 ft. 9 ins. would seem adequate."[12]

There was yet another makeshift scheme during the period before the acquisition of land at Riccarton had been authorised, and it was just a question of waiting for the buildings to be completed: rooms were to be provided for staff in the attic of one of the buildings, at that stage being used as a " 'junk' store-room." According to Calvert, (who by this time had been appointed to the Chair of Civil Engineering and had accepted the Deanship), "to build 4 to 5 staff rooms would be a comparatively easy matter. The addition of two dormer windows and light partitions (half glazed) is all that would be required, with the possible exception of alternative egress in case of fire."[13] By this time the whole building was a potential death-trap, not just in the

*see p. 175

possibility of fire but earthquake as well. Various safety regulations issuing from the Government or the M.E.D. were by-passed, or were not put into operation because of the crippling expenses involved. Students were advised always to wear sandshoes to provide them with an effective insulation in the event of an electric shock, and to keep their hands in their pockets in case they were tempted to fiddle with the switches. In 1944, Powell requested a switch to connect the School's internal telephone system to the G.P.O. external lines. Inspection was followed by the outraged comment: "it has come under notice that the private link connecting the 7 magneto telephones is wired with bell wire . . . the insulation resistance to earth is practically zero."[14] Teaching or studying at the School in those early years was undoubtedly not without its hazardous side. Conditions, which today would seem excessively casual, were nothing exceptional in those days.

The office staff likewise suffered greatly; extensions to the School in 1913 meant that the room occupied by the typist now looked out on a forbidding stone wall. Miss McIntyre had been appointed in 1920. Her room received no direct light at all and Powell pleaded strenuously but in vain that at least the room could be brightened by a fresh coat of paint, as it was most gloomy and depressing for the young woman to spend all her working hours in such drab and comfortless surroundings.

The lack of space affected not only staff accommodation, but as the numbers of aspiring engineering graduates increased, especially after the war, laboratories became too small to accommodate the larger classes. Accordingly, laboratory sessions had to be duplicated, then triplicated and finally in 1958, the year before the move to Ilam, one of the classes had to be repeated five times. (Of course the same situation obtained in the Science Faculty which could gratefully spread out a little once the School moved out.) Before the move, the Chemical Engineering Department (established in 1944 under the auspices of the Chemistry Department)* occupied a basement in the already overcrowded Chemistry Department. It says a lot for the patience and forebearance of the staff at the time that they endured such conditions, but it is not surprising that there was no energy nor enthusiasm for research projects after five identical laboratory classes in a week. And it was not even as if these laboratory classes had a workable number of students; they inevitably contained the full possible quota and more. Steele related sadly, and without a tinge of criticism despite the poor relationship

*see p. 166

that existed between him and A.Lush,* that all Lush could achieve because of the excessive numbers in a problems class was to present the problem one week and then provide the solution the next, hoping that in the meantime the students would have tackled it themselves. It was quite impossible to check each individual student's work.

And it was not just the makeshift re-arrangement and clutter of small rooms, the blocking off of a staircase there and the partitioning of a corner somewhere else, that contributed to the gloomy presentation of the building. Miss McIntyre's room was not the only one which needed painting. There was a scathing letter from T.Pollard, Acting-Professor of the Electrical Engineering Department in 1947 after Powell's retirement. The quality of light in the room where his students were supposed to be working on drawing and design was intolerable, with or without artificial illumination. As he claimed, "the whole idea of this class becomes somewhat of a farce," unless the students could carry out "actual drawing."[15] It was suggested that if fluorescent lighting could not be installed — as indeed it could not, as it had to be imported from the States and University requirements were not classified as deserving of top priority in those post-War years — then surely at least together with higher wattage lamps, the walls and ceiling of the room could be painted in order that they would reflect light instead of absorbing it. It turned out that many of the rooms within the School had not been painted for at least 30 years. R.D.Neale, a lecturer in Electrical Engineering, was asked by the Council in 1943 to present a report on the quality of lighting within the College. Many of the lecture rooms had an illumination value of two to three foot candles and therefore came into the category of illumination which was suitable for churches, offering merely "a dim religious light."[16] Calvert managed to organise better illumination for the surveying room but the other rooms continued in their dreary and inadequately lit state. It is a wonder that more staff did not suffer from eye difficulties when we consider the conditions under which they worked.

If lighting during these years was inadequate so was heating. The original gas fires or even open fires had been replaced eventually by a College heating system which was partly devised by Steele, following plans and specifications previously laid down by Scott. Steele likewise

*A graduate of the School, B.E.(Elect.) 1916, Lush had worked both in New Zealand and the United Kingdom before taking up his appointment in the Mechanical Engineering Department at Canterbury in 1936.

supervised its installation. Ultimately, of course, the boilers for this heating system became derelict and developed cracks, thus becoming dangerous. (Stoking them had always been a risky business because the boiler room was below ground and tended to flood. As well, access was down a dangerously steep flight of steps. As the room was ill ventilated, the poor fireman had even more unsatisfactory working conditions than the typist.) When the cracks first became obvious, R.J.Rastrick,* after ten years as Engineer to Vale and Co., had recently been appointed Head of the Mechanical Engineering Department. The burden of responsibility, therefore, fell on him. He advised replacement boilers, which had to be imported as neither Andersons nor Vale and Co. could provide them. Importing meant delays and money and over the next eight years the heating system of the entire College became increasingly inadequate until it was defunct. The Department of Education at one stage questioned Rastrick's estimates which brought forth a very astringent letter from him, reminding the Council how much money it was saving by his giving it the benefit of his services and advice, and protesting that his professional opinion should be treated so scurvily. "In conclusion," he wrote, "I would like to state that our proposals were based on seasoned experience of our requirements and, frankly, I object to my expert opinion being questioned on such flimsy and irrelevant grounds."[17] As was customary in such situations, the Registrar had to write to the Director of Education and express, as tactfully as possible, his and Rastrick's dissatisfaction. (Rastrick's written style is admirable; however, he did not insist that his students attain such fluency. In 1950 he was asked to comment on whether the literacy level of engineering students had deteriorated. Provided they were capable of compiling adequate reports he was satisfied; there was no need for them to produce "sonorous Johnsonian English."[18])

Meanwhile the College froze; *Canta* was full of complaints under such cryptic headlines as "Life begins at 40.°" but attending lectures and long laboratory classes in cold stone buildings on frosty mornings was not reasonable, and the Council did all it could to hasten the Department of Education's procrastinatory decisions. The whole situation dragged on interminably with many students spending virtually their

---

*Rastrick, Ronald John Whittington, 1903- ; educ CU (BE), 1929; after graduating he took up a British Thomson Houston Exhibition in the U.K.; lecturer-demonstrator CU 1930-1935; engineer to Vale and Co., ChCh; 1935-46; prof mech eng 1946-68. He was responsible for much of the planning of the Ilam School.

whole time at the College in oppressively cold conditions. (The custom of running electric heaters off the light sockets was not confined to the student hostels. The M.E.D. reported that it was a common practice within the College as a whole. As the College had been granted special rates for lighting, the M.E.D. did not appreciate a situation which made a mockery of this concession, apart from the dangers inherent in overloading the electric circuits.) It was not until 1955, eight years after it had been first reported that the boilers were beginning to cause problems, that the College once more had a workable system.

The lack of aggressive vitality after Scott's retirement, which kept the Board and the Council concerned over the years as to whether or not it would be advisable to appoint a Director of the School, also meant that requests for increased accommodation or complaints because of inadequacy from whichever of the two Professors was currently Dean carried little weight. Gone were the days when Scott was able to insist that his windows be double-glazed to muffle the noise of concrete chipping when the Library was being erected. Although he had to express his indignation in more than one letter, once he had made it clear that the noise made it impossible for him to carry out his teaching duties, the Board of Governors had to accede to his request. By contrast, Powell was particularly concerned that the Board were overlooking the very great need of the School for extra space when the old Boys' High School building became available for College purposes in 1926. He claimed that Scott had reached an agreement with the Board as to how much space would be allocated to the School and that the Board was reneging on that agreement. All his protests and representations, however, did no good; there were, in the Board's eyes, other equally deserving claimants and Powell and the School were expected to be grateful for the portion meted out to them. All they acquired was the west end of the old school. Inevitably they were disappointed, especially as any plans for putting up new buildings had to be shelved as the depression years began to bite.

Upkeep and maintenance of the buildings, both inside and out, was at a minimum during these years and afterwards. Stonework and mouldings were allowed to crumble away, disastrous cracks appeared in the outer walls, skylights needed reglazing, and guttering was dilapidated. Against all these serious, and largely ignored, items of upkeep, the need to replace wirenetting over the windows facing on to the Boys' High School playground to prevent footballs coming through strikes a lighter note. The fives court always presented a bother as persons getting on the roof of the battery room to retrieve their balls did various unspecified damages and Powell wrote peevishly protesting more than once about the inconvenience caused. On the other hand the School could shamelessly encroach on the High School's territory, as a letter from the Headmistress of Girls' High School, Miss Gibson revealed. Would the School of Engineering, she politely asked on 11 February 1926, be so kind as to remove the wood it was currently storing in the Boys' High swimming pool, as her pupils wanted to make use of the pool for regular swimming at the beginning of the coming term.

Later, in 1953, there was a disastrous fire in the old Boys' High School building, as it continued to be called, and immensely expensive and recent surveying equipment narrowly escaped incineration. Other damage was caused by the water which extinguished the fire and then, because the building was left unroofed for some time pending repairs, there was further water damage from rain. Apparently the building was subsequently invaded by most impressive dry rot, the giant fungus occupying whole drawers in one of the engineering rooms. The Boys' High fire was serious, but from time to time there had been less serious fires reported always with gratitude at the prompt action of the man who had spotted them and prevented their getting out of control. Considering the jerry-built constructions, the inadequate and frequently drastically overloaded wiring, it is truly remarkable that fires were not more frequent.

CHAPTER THREE

Staffing Difficulties: Professors, 1923-1959

One of the most difficult problems that beset the School of Engineering during its first seventy-five years was maintaining adequate staffing levels. Some of these problems were common to all University departments; others turned on special problems peculiar to the engineering profession.

If it was always difficult to attract appropriate men to take up positions at the School below the rank of professor, appointing a suitable man to a chair was by no means straightforward. Campbell resigned in 1929. Because his resignation led to the re-examination of whether the new professor should also be the Director of the School,* there was a considerable delay in finalising the appointment to the chair. As a result, Calvert, who had been an active and enthusiastic staff member since 1921, was offered £600 to take on the role of Acting-Professor for the year 1929. (Initially, the amount offered was lower and it was only raised to the £600 after he had strongly protested.) Calvert himself was on the short list for the position but it went finally to J.E.L.Cull, who had been a demonstrator, in the Scott years, between 1902 and 1908.

Wood, who never let well alone, was incensed at the appointment. He held it against Cull that he was a "product of two of our local educational institutions" — Cull had attended Boys' High School, as well as being a graduate of Canterbury College. In addition, "Mr Cull has spent a great part of his active life in the service of the Public Works Department, a sphere of activity in which one would not expect to find embryo professors." But Cull's greatest sin was that he had an excellent reference from Professor Scott, "especially striking in the exuberance of its laudatory expressions." Wood was of the opinion that Scott had "completely misled the Board . . . by his excessive laudation" of one of his staff members at the time of his retirement. His final

*see p. 103

reason for opposing Cull's appointment was that at 51, he was too old. "University Colleges throughout the Empire aim at catching their professors young, about the age of 30."[1] While we can admire Wood's persistence, we have to recognise that the Board made a pragmatic appointment. It really had no option unless it were to accept that it would defer the appointment and allow Calvert to continue in the role of Acting-Professor indefinitely. Criticism of the quality of appointment to staff positions throughout the years tends to ignore the realities of the situation. As £850 was the maximum the Board could offer and as many creative engineers would not be attracted to academia, even if lured by a higher salary commensurate with what they could earn in private practice, the range of applications received at any one time was severely limited. If Wood was right and Cull was not an ideal appointment, he was the best the circumstances offered. Certainly it remains true that the Civil Engineering Department under his leadership went from strength to strength.

Cull held the position of Professor of Civil Engineering until 1941, when he was compelled to retire early because of ill-health. Once again, Calvert was asked to accept the role of Acting-Professor. The position was advertised — Professor of Civil Engineering who would also be Director of the School — but the Council quickly perceived that it could not hope to make a satisfactory appointment while the war was so drastically reducing the field of applicants. Calvert, therefore, remained as Acting-Head until 1946. Once the war was over, advertisements were again placed in leading engineering journals and newspapers in the United Kingdom and Australasia. On the recommendation of a selection committee set up in Great Britain by the Universities Bureau of the British Empire, Major Daymond, of the University of Liverpool, was offered the chair. He accepted but finally was unable to take up the appointment because, while he could arrange priority shipping for himself, it proved impossible to make any definite arrangement for his wife and family to accompany him. The other two strong contenders for the chair were Colonel W.G.Morrison* and Calvert himself. The Council decided to interview both men, but in the event, Morrison declared himself out of the running by stating that he was only willing to consider the appointment on the terms of the original advertisement: that the

*Morrison, Gordon W., Educ CC, B.E.(Civil), 1926; worked for five years as "Contractors Agent" for Wests Piling Company, London; war service in India; partner in firm Campbell and Morrison; Hon D.Sc., 1973.

position would be that of the Director of the School. As he said:

> Salary is not the issue as far as I am concerned for either alternative. It is a question of human and scientific interest and opportunity; the Directorship basis, implying as it would, sufficient staff to free the Director of too many routine duties, would allow time for the consideration of forward policy and the development of research.[2]

As the Council had rescinded the previous decision to appoint a Director, it was hardly in a position to modify the terms of appointment to suit Morrison. And while his practical experience was impressive, doubts had already been expressed at his not having had any teaching experience. On the other hand, Calvert certainly had teaching experience — in 1934, Cull recommended that he should be granted an increase in salary on the grounds that "his value increases yearly as he gains in knowledge and experience"[3] — but he had no degree, holding only the associateship of the School. However, in this respect he was in good company. Campbell, in his testimonial for Calvert even reminded the Council that "the late Professor R.J.Scott"[4] had also been without a degree. Consistent to its bias of preferring overseas candidates to local men, the interviewing committee for the Universities Bureau expressed concern on the grounds that it "might be regarded as instituting a 'caretaker' period of ten years or so . . . cannot think anything impressive would result . . . at 54 he has shot his bolt."[5] Yet, A.H.Tocker,* Rector of the College, and another of Calvert's referees, stated that Calvert was greatly to be commended for encouraging research against all the odds of inadequate funding, insufficient staff and an unsuitable building. He even partly attributed the popularity of the Civil Engineering course to Calvert's enthusiasm and energy. In the end, such qualities, and his familiarity with the conditions of civil engineering within New Zealand, far outweighed Calvert's lack of academic accomplishments and he was appointed.

Calvert held the chair until 1950 when he resigned in order that his replacement could be at the centre of the planning for the move to Ilam, which was, in the end, delayed for nearly a decade. H.J.Hopkins, who was appointed in his place, came from Western Australia. By this time, the Council had had considerable experience of the kinds of

*Tocker, Albert Hamilton, 1884-1964; Educ VUC (MA), Tocker served in Ist NZEF, was asst lect in econ 1921-5, prof 1926-50, Rector 1943-8, m Sen UNZ 1944-9.

difficulties it might face by bringing in an outsider, as is apparent from the careful letter the Rector of Canterbury College, Dr Hulme,* wrote to the Western Australian Vice-Chancellor, G.A.Currie, asking whether Hopkins was a tough enough character to cope with one or two "disappointed senior lecturers."[6] The answer was reassuring: "Hopkins had a good enough opinion of himself not to be bluffed by any disappointed senior lecturers." It was also stated in Hopkins's favour that he had "a most excellent wife."[7] The redoubtable Hopkins was certainly strong enough. From the very beginning his letters to the Council revealed a man imbued with considerable confidence and certainty of direction. He was undoubtedly a force to be reckoned with.

Neither of the two senior lecturers — P.J.Alley** or D.Bruce Smith — had been considered for the chair; the short list had contained, apart from Hopkins, Major Daymond, who was by this time in South Africa, and L.Holmes. When Calvert retired, Alley was the most senior member of the Civil Engineering Department and could, therefore, have expected to be Acting-Professor until the appointment was secured. With an unusual break in precedent, the Council requested Rastrick, the Dean, to undertake the headship of Civil Engineering, which constituted a decisive snub to Alley and certainly said very little for his chances of gaining the chair. His protest at this treatment was of no avail. His next move was to resign from the department, which would have left the Civil Engineering Department in a most precarious position. The Council's reaction was neutral; Alley was to be informed "in due course of the Council's decision re the date from which [he would] be released."[8] However, a week later, he was requested to attend "an interview with the Staff Committee,"[9] the upshot of which was that he withdrew his resignation.

Not only was Hopkins "not bluffed" by his disgruntled senior lecturers but during his years in the chair he was very loyal and good-humoured in his support of Alley. (Smith resigned in 1951, regretting that he had had such a brief spell of working under Hopkins.) In 1953, Alley applied

---

*Hulme, Henry Rainsford, 1908; educ Manchester GS, Cambridge and Leipzig U; Fellow Gonville and Caius College, 1831-37; lect. maths Liverpool U 1936-38; chief asst. Royal Obs., Greenwich 1936-39; Admiralty research 1940-45; director naval research 1945; sci adviser Air Ministry 1945-48; Rector CUC 1948-54; subsequently chief nuclear research, Atomic Weapons Research Establishment, Aldermaston.

**Alley, Philip John, 1901-78; educ CC, B.E.(Civil) 1925. Alley had worked for the P.W.D. before his appointment. He had a strong interest in stabilising soils both for roading and for house building, and under his guidance a number of soil-cement houses were built in Christchurch and elsewhere.

for leave of absence for five months and his application was given Professorial Board support. Hopkins was invited to attend a standing committee convened to discuss the question of Alley's refresher leave, and obviously protested vigorously when the Council refused it. Alley, himself, indignantly requested that the Council inform him of the reasons for its refusal. The reasons were: "the Council does not consider that Mr.P.J.Alley's services to the College warrant the expenditure of any moneys on Refresher Leave."[10] Alley, who was a regular correspondent to the newspapers on questions of New Zealand's relations with communist China, made this whole issue public in the *Press*. For this, he was criticised in *Canta*, which, nonetheless, was outspoken in its condemnation of the Council's behaviour. *Canta* assumed that the Council's real reasons for refusing the refresher leave related to the fact that Alley, after attending the Third International Conference of Soil Mechanics and Foundation Engineering in Zurich, had purposed visiting engineering works in Russia and China on his return trip:

> If, as is by no means impossible, a dislike of Mr Alley's political views have (sic) in any way affected the decision of the College Council in refusing to grant him refresher leave [especially after he had received Professorial Board concurrence], then the College Council is regarded as having departed from the generally accepted principles of academic freedom — the principles which allow a university teacher to hold and express views, economic, political or religious, disregardless (sic) of whether they conform to those of the political, economic, or religious status quo.[11]

Alley was a man highly knowledgeable and enthusiastic about his subject, but with no ease of manner. His inability to assess what would be appropriate in a social situation, can be seen by his method of entertaining a group of first year students during Orientation week in 1956, few, if any of whom were engineers. The highlight of the evening was a very scratched and slow-moving film of the collapse of the Tacoma Narrows Bridge.

When Steele retired in 1946, his replacement was once again a local man: R.J.Rastrick, returning to the academic world after his ten years with Vale and Co. J.L.Thomson, from the University of London was offered the position, but he had accepted another appointment during the long delay while the Council deliberated. Once again the Universities

Bureau were suspicious of the local candidate, questioning Rastrick's motives for deciding to return to academia and hoping that his application was "not based on fear of industrial depression."[12] Rastrick had strong local support, his knowledge of the industrial scene in New Zealand being greatly in his favour. It was firmly believed that "he would be able to put the teaching of mechanical engineering on a modern and worthwhile footing."[13] As, out of all the departments, that of Mechanical Engineering had deteriorated the most in the years since Scott had retired, Rastrick faced a difficult challenge in attempting to revitalise it.

The final professorial replacement of these years came at the time of Powell's retirement. N.M.MacElwee* from Birmingham was appointed to the chair of the Electrical Engineering and was immediately at daggers drawn with his senior lecturer, T. Pollard. (That story belongs to a later discussion about the War years and the establishment of Electronics.)** In retrospect, we can see that the appointment to chairs of native New Zealanders, ex-graduates of the School and familiar with its ways conferred both a positive and a negative blessing. Such appointments scarcely disturbed the status quo at a time when the School was struggling to improve its situation and faced with severe difficulties. On the other hand, these appointments continued the Scott tradition far beyond the time when it was viable or to the School's advantage, and probably contributed to the growing dissatisfaction which surrounded the School and its greatly diminished effectiveness during these years.

*MacElwee, Norman MacLeod, 1899-1976; educ Glasgow, B.Sc.; acting-prof Elect Eng Birmingham 1939-47; prof CC, 1947-65. MacElwee had wide research experience in both the machine and radio branches of electrical engineering and was engaged during WWII on problems related to servo control and microwave radar.

**see p. 175

CHAPTER FOUR

Staffing Difficulties: Junior Staff, 1923-1959

> There is another point which, to my mind, is a very important one, viewed in the light of experience of the organisation of various engineering undertakings. It is this: there is too great a gap between the professor, as head of a department, and myself, as a lecturer. This wide interval is probably an unwanted legacy of a defect of Professor Scott's regime, and even in his day, it must have been a weakness . . . this interval might now be likened to that between a major and an n.c.o. in the army. A major without any captain or lieutenants to whom to delegate any authority is at a major disadvantage.[1]

The term "god-professor" has been coined to describe the very difficulties Lush was outlining in this persuasive, but futile attempt to have his salary and status in the Mechanical Engineering Department upgraded. It encapsulates many of the problems faced by small departments when one man was virtually in sole charge and any other staff members were seen to be entirely subordinate. The extreme hierarchical system which obtained in Scott's early days in the School of Engineering ensured that all power, prestige and responsibility rested with him, and A.F.Morrison was in the invidious position of merely following instruction. Every time Scott was absent on leave, Morrison endeavoured, in vain, to persuade the Board that he was deserving of an increase in salary. (He was always an opportunist; he even entreated the Board to establish a mini-department of Applied Mechanics after Scott's retirement, with himself as Associate-Professor. Considering Scott's attitude towards him over the years, he was lucky even to become a lecturer at this time.) The Board always waited until Scott's return before making a decision, feeling that Scott's recommendation was necessary. But a demonstrator was clearly regarded as a second class citizen, and it is not surprising that Morrison chafed against the situation. But appeals to have his position redefined were in vain; Scott was intransigent. When Morrison eventually retired in 1936, the most significant comments which were passed about his career concerned the fact that he had been a member of staff for so many years, almost from the very beginning of the School. As he had been one of the earliest engineering students, he had certainly an extensive knowledge of the School's development. "Baldy", as the students affectionately called him was greatly valued by them for his approachability and the keen interest he took in their progress.

Initially Scott was able to run "his" School with only Morrison's assistance. But, as the years passed and the range of courses offered by the School extended, it became necessary to appoint lecturers. The first lecturer in Electrical Engineering, Craddock, found that his ambitions were thwarted by Scott's unassailable power and conflict developed.* Other appointments were less abrasive; Steele and Powell were to continue at the School for the remainder of their working lives. But until the numbers of degree students equalled or outweighed the numbers of evening students it was impossible to justify the appointment of a man to lecturer status. It is uncertain whether Powell and Steele retained the hard and fast demarcation line between demonstrator and lecturer which Scott had initiated; certainly Rastrick, appointed demonstrator in 1929, was given the status of lecturer-demonstrator the following year, and Grundy, who had been appointed from the United Kingdom in 1923, gave as one of his complaints against the Department at the time of his resignation in 1929, that, although he was designated lecturer, he had failed to receive this title in the School of Engineering prospectus. (Grundy, Pollard, and Calvert had all been promoted to lecturer status in 1925.) This would suggest that there was still an important distinction between the two ranks. But even when most of the staff under the, by now, three engineering professors, were of the rank of lecturer, organisation of the different departments was far from democratic. The A.U.T., which was established in 1934, worked to introduce the position of senior lecturer, rightly of the opinion that if the Universities were to attract good men and women, there needed to be a clearly defined ladder of salary and status. Very few people were willing to moulder in the same menial position for the whole of their working careers without some guarantee of promotion and recognition. Few men would

*see p. 67

have regarded their time with the School the way W.R.Robinson evidently did. He retired in 1935 after 25 years of faithful service. In his retirement speech, Robinson declared: "the farewells accorded me cause me to believe that the part of my life spent in honourable service to you has been in no wise wasted or useless. If so, I am pleased, but take no credit to myself."[2] Confidential staff reports over the years consistently referred to the "excellent" quality of his teaching, and student tributes in *Canta* spoke appreciatively of his enquiring mind.

Throughout these years, however, the main problem was the very low salaries offered to prospective Engineering School staff, whether they were demonstrators or lecturers. The position of demonstrator or even demonstrator-lecturer was not such as would be long endured by an ambitious man, seeking improved status and salary. When I.Hayman left in 1936 to take up a better paid position with the Seddon Technical School, Powell was incensed that the School was losing staff because the salaries it offered were so uncompetitive. "I think it my duty to bring to the notice of the Council, that the starting salary for a junior position in a Technical School is sufficient to attract a member of our staff who has spent 11 years with us."[3]

Changes in the outside world had also affected the School and its ability to attract staff. The word engineer had lost its dubious connotations. Thus, at Campbell's farewell function in 1929, the Rector, Dr Hight, could condescendingly remind his audience that "Professor Campbell was an engineer . . . a word which formerly had a questionable significance, but which today stood for honesty, accuracy, and rectitude."[4] The New Zealand Institution of Engineers worked constantly to upgrade the profession, actively encouraging the 1924 Engineers Registration Act. The offshoot of this, successful, attempt to raise the status of the engineering profession was that engineers could command higher salaries. Unfortunately this had a backwash effect on the School, because it did not have the financial resources to compete with industrial firms or the P.W.D. and so could not attract to the School men of sufficiently high calibre to satisfy its needs. Surveying the possible field of applicants to replace Rastrick, when he took up his position with Vale and Co. in 1936, Steele was very gloomy: "owing to the greater number of engineering positions available this year I do not think there are likely to be many suitable applicants."[5] When W.S.Wright was appointed to replace Hayman, he was described as "the only one suitable out of a poor lot;" he was "not impressive as a teacher," being "dreamy and

The School's staff in 1946. Back: I.F.D. Wilson, N.E. Hewitt, T.M. Stanton, P.J.Alley. Middle row: M.A.A. Allfrey, B.T.W Owen, I.L. Holmes, W.J. Syme, W.S. Wright, A. Lush, A.H. Simpson. Front row: R.J. Rastrick, R.D. Neale, P.H. Powell, T.R. Pollard, G.G. Calvert.

incapable of hurrying."[6] (When Wright retired in 1944, he was described as "a faithful servant of the College.)"[7] Later, in 1943, the N.Z.I.E. spawned a break-away group, the Professional Engineers Association, with the sole aim of acting as watchdogs on the status and salaries of professional engineers. On several occasions a representative of the P.E.A. wrote courteously to the Council protesting at the low salary advertised for an engineering lectureship. The Council always wrote back, equally courteously, although without making any adjustment to the salary.

It has been said that it is possible to judge the quality of a Head of a Department by the kinds of appointments he makes. In ideal conditions, this may well be true; but no one would pretend that the conditions which obtained prior to the School's move to Ilam were ideal, or even reasonable when it came to appointing staff. The need to hold out more generous salaries to attract better applicants was a constant refrain in letter after letter to the Council. Pollard even went so far as to indicate that he would not trust an applicant who would accept £400 a year. "We could get nothing better than a radio serviceman, who would be quite unsuitable for the job."[8] Rastrick, in similar vein, declared: "I am frankly very disappointed with the response we have had as no application can be considered as really outstanding, and I am faced with the appointment of men of moderate qualifications only."[9] The choice confronting him was quite clear-cut. Either he appointed a mediocre man to the position or he re-advertised in the hopes that the fresh applicants would be of higher quality. A delay in appointment would severely burden the existing staff. Consistent understaffing caused a "policy of distributed neglect."[10] The problem was exacerbated after the war with the overcrowding caused by the influx of returned servicemen and others who were responding to the boom in technology. By 1947, all three professors, Calvert, Rastrick and Pollard (during the year he was Acting-Professor pending the appointment of a man to replace Powell) concurred in their dismay at the situation.

> We wish further to state that we are very definitely of the opinion that, entirely due to the . . . staffing shortage, the standard of instruction to our Engineering students has been lowered, and the students at the College at the present time are not being as well prepared for their profession as they have been in the past. We are

> concerned that the good name which the School has achieved in past years may be injured.[11]

Their fears were rather derisory considering the already growing dissatisfaction at the quality of engineering education offered by the School. By 1950, the Rector had taken a hand in the game. Rastrick was currently Dean. At the same time he was acting as Head of the Civil Engineering Department. A letter from Rastrick to Hulme outlined the "parlous position" the School was in, greatly aggravated by the presence in Christchurch of one of the Commissioners from the New South Wales Snowy River project on the look-out for engineers to assist with the development of hydro-electricity in that region. Every member of the Civil Engineering staff was dissatisfied or had already resigned. Other departments, although they were faring better, had no cause for complacency. Consequently Hulme wrote to the Minister of Education maintaining:

> . . . it is essential that the Engineers be put on a special scale as are the Medicals and Dentals in Otago. Only in this way can we keep the School together and attract suitable staff to fill future vacancies. If something like this is not done, it seems clear that we may have to seriously consider closing at least the Department of Civil Engineering.[12]

A differential scale of salaries for engineers never became a reality. (In itself, it was a sensitive issue. A differential scale affecting all senior teachers of the College had been most reluctantly accepted by the A.U.T. in 1949 "under protest as a temporary expedient." Although "there is no evidence that [this] improved the quality of the staff," it did leave "bad blood)"[13] The School of Engineering had to continue competing on most unequal terms with industry, not just within New Zealand, but also in Australia and further abroad. Many of the best Canterbury graduates chose anyway to try their luck overseas. It is notable that out of the nine Engineering Scholarships which had been awarded by 1929, only two of the recipients were working in New Zealand. (This had always been an argument against awarding such scholarships in the first place.) New Zealand was competing in a world-wide market, a market which could offer far more attractive incentives to interested applicants. It was not just that Australia and Great Britain gave higher salaries. Overseas candidates faced further difficulties after the war, with delays in procuring priority passages, especially for dependent relatives — one prospective appointee turned down a position because he did not want to leave an elderly parent behind. The extreme housing shortage which meant that new arrivals might have to face an unconscionable time of boarding before they could find suitable accommodation was an additional disincentive. Canterbury University College's School of Engineering, even when it was designated the National School of Engineering was a laggard in any race to attract staff against such odds.

The establishment of the position of senior lecturer did not eliminate conflicts and personal jealousies; in fact they were exacerbated. Promotion occurred as a result of the confidential annual staff reports from Heads of Departments; these reports were assessed by the Rector who then reported his recommendations to the Council. Where, as in the case of Steele and Lush, there appears to have been a most unsatisfactory relationship between a professor and a lecturer, then such promotion could be deferred indefinitely. Lush certainly felt this very strongly. As the years went past and he was neither promoted nor granted a higher income, his appeals to the Council that it reconsider the justice of his claims became more shrill and insistent. What particularly stuck in his gullet was that N.E.Hewitt, who had been appointed in 1940, with fewer years of service to the School, and without growing children to educate, had a senior lectureship conferred on him. (The staff report in which Steele recommended that Hewitt be promoted, warned that he would be likely to seek work elsewhere if he were not given some incentive to stay with the School.) Steele found Lush's work not "satisfactory,"[14] and obviously not worthy of greater recognition. However it was rare that any member of his staff called forth a stronger commendation from him; while this may reflect on the quality of the staff of the Mechanical Engineering department, it may also be a feature of Steele's unenthusiastic nature. One year, 1945, Steele's assessment even read cryptically: "The quality of Mr Lush's work does not appear to have altered this year."[15] He was obviously irritated by Lush who was constantly, like Morrison before him, trying to circumvent College procedure by appealing directly to the Council and by-passing Steele. The difficulties between the two men were constantly aggravated by Lush's requests to be granted permission to undertake holiday work for the Public Works Department, and here again Lush unnecessarily irritated his superior by writing direct to the Council and ignoring the customary procedure. On another occasion he requested permission for three months' work when the rest of the staff would receive only a six weeks' break. Steele grudgingly gave his recommendation, but commented to the Council that such preferential treatment could only lead to dissatisfaction amongst other members of the staff. Reports that Lush's work was not satisfactory or had not changed any from the previous year give very little indication of the real problem, although one statement from the Rector, in 1939, revealed that he was one of those unfortunate men in any department who fail to discipline classes, which would suggest that for all his obvious interest and enthusiasm for his subject he was not an inspired teacher. Eventually he became sufficiently desperate at the Council's repeatedly ignoring his appeals, that he asked to be allowed to attend a Council meeting in person in order to present his case. Surprisingly, this was permitted, and possibly as a result of A.U.T. intervention he became at last a senior lecturer in 1949. (Lush obviously never ceased to hold a grudge against the Department for the cavalier way in which he had been treated and several years after he had been compulsorily retired, at the age of 65, he was again soliciting the Council for some kind of retirement grant on the grounds that his financial position had been seriously imperilled by his not being promoted at the right time. His claim was dismissed.)

Out of the staff of the Mechanical Engineering Department in those early years, Lush was not the only member to feel disgruntled. Bamford, who had been the butt of many of Scott's jokes, remained in the department but retired quite suddenly at the same time as Steele, without having given any previous warning of his intention. He had likewise battled to receive more recognition for his services, although in his case he had the complete good-will of Steele. Bamford's position in the Department was Testing Engineer, a description which baffled the Council on several occasions. Steele was requested to explain exactly what it was that Mr Bamford did. As most of his time was spent on the very demanding testing programme which the Mechanical Engineering Department was forced to undertake, his teaching was limited to supervising laboratories and occasional lecturing, ie, reading someone else's lectures. Like Lush, and considering his much longer time with the School, with even more justification, he greatly resented it when Hewitt was promoted to senior lecturer status. It was more humiliating for him, in that he had actually taught Hewitt. Steele was sympathetic to Bamford's feelings in the matter but could not see his way clear to promote two men at once. Eventually Bamford, too, felt sufficiently strongly to write direct to the Council. As all communications with the Council had to be directed through the Head of the Department — Scott had been outraged when this procedure had accidentally not been observed — he was rebuked for not observing the protocol. He received no joy and shortly afterwards resigned.

Lush's and Bamford's predicament illustrates something of the difficulties faced by long term staff before the days when the ladder of promotion was firmly in place and when staff numbers were still so small that more than one senior lecturer was a luxury the Council could not afford. It is perhaps not a coincidence that they were both in the Mechanical Engineering Department. Bamford especially, having been appointed in 1916, had borne the brunt of the depression years, and the whole damaging situation when the turn-over of staff was so rapid that no sooner had one man been appointed and learnt the ropes than he was off and a new man had to be initiated into the running of the different courses. The low salary, low prestige and the excessively long hours caused by the two-fold function of the School, as a University and a Technical School catering to evening students, is sufficient explanation of why men stayed the course for so few years. The School of Engineering was bedevilled for many years by the unsatisfactory salaries offered to its junior staff, which did not compete with what energetic engineers could earn elsewhere. And even if those who applied were fully competent as practising engineers, there was no guarantee that they had either a gift or a vocation for teaching.

When the School of Engineering was established, the initial terms of appointment to the staff expressly stated that no work was to be undertaken outside the School. Apart from serving on various Government Committees of enquiry, Scott obviously also won permission from the Board to undertake consultancy work. He himself referred to his having advised the Canterbury Hospital Board when a new heating system was installed, although a careful search through the Hospital records makes no mention of his name. Other staff during the Scott era had to accept the prohibitory clause but, as time went on, it became necessary for the Board to accept a modification. Calvert was probably the testing case. In 1922, E. Hitchcock, the Chief Electrical Engineer of the M.E.D. requested Calvert's services in "connection with investigation into the hydro-electric possibilities of the Waimakariri River." Calvert applied

T.R. Pollard

N.M.MacElwee

War-time work on radar aerials in the School in 1944.

to the Board for permission but this was not granted. It took a special letter from Hitchcock regretting his "non-availability"[16] before the Board granted the concession. Even so, it stated firmly that it did not want this to create a precedent. This is rather derisory in the light of the Board's unashamed exploitation of the engineers on its staff. The same year, 1922, Campbell confronted the Board with his opinion of the unsatisfactoriness of the Board's regulation about not undertaking consultancy work. He declared:

> I have been asked on several occasions to make or check calculations for structural design in steel and reinforced concrete, in both of which I have had special training, and I consider that the carrying out of this work would keep me in touch with present day practice and prices; would give me a certain amount of useful remunerative work during vacations; and could in no way interfere with my College duties.
>
> Professor Scott has always enjoyed the privilege of doing private consultancy work, some of the results of which have been embodied in his lectures and afford excellent examples of the principles involved.[17]

The Board relented and thereafter, until his resignation, Campbell submitted annual reports of the private consultancy work which he, often with Calvert's assistance, had undertaken. Neither Steele nor Powell ever contested the Board's ruling nor submitted details of outside work. Either civil engineers were more in demand, or the pressure of administering the School year by year as well as undertaking all the School's testing work sapped their energies. Although the Board became more lenient towards requests for the staff to engage in outside work, the clause banning new appointees from accepting such work remained as one of the terms of contract. Rastrick, on his appointment as a demonstrator in 1929, protested at the unfairness and was reassured that the clause was at the discretion of the Board who would regard any requests for permission to undertake consultancy sympathetically. True to character, at the time of his appointment to the Chair of Mechanical Engineering in 1946, Rastrick again queried this clause in the terms of his appointment and received an almost identical answer. As the depression pressed more heavily, lower paid members of the staff greatly needed the extra money they could earn, wherever possible, from outside practice. The other argument, that it kept an engineer more in touch with the latest developments in his profession remained constantly valid.

CHAPTER FIVE

The School During the Depression

Having lost its forward momentum at the time of Scott's retirement, the School of Engineering, like the rest of the country had to endure the years of the depression. Canterbury College itself was in a perilous situation. The very factors which had given it prosperity at the time of its foundation, namely the endowment lands ceded to it by the provincial government, now militated against it. The College was landlord to numerous tenant farmers throughout Canterbury and as the depression settled over the land, the farmers received less and less for their primary products. Rent fell into arrears. Pleading letters arrived at the College and the Board really had no option but to deal as generously as possible with its troubled tenants whose plight was so overwhelming. The pathos of the letters was almost unbearable, many of the farmers struggling not only with financial disaster but with the problem of articulating their difficulties so that the Board of Governors would regard them with leniency and forbearance. Each case was examined on its individual merits; considerable licence and extension of payment time were granted. But the College itself inevitably suffered, especially as the Government, pressed on all sides with urgent and legitimate economic claims, was able to suggest that its grant to the College should be less because of the existence of these very endowment lands which were causing the College such embarrassment. Whereas in 1928 the arrears of rent were at a level of £2,460, by 1933 they had reached £20,227.

"Fearing the growing power of the Education Department, the Board was reluctant to make importunate appeals for State support, and in any event the State, immersed in financial problems of its own rarely listened."[1] By this time, however, the School of Engineering was in no position to refuse Government assistance for what was proudly acknowledged to be a "national" institution. Thus, while the School undoubtedly endured considerable hardship during the depression years, it was not as financially straightened as the College proper. In 1933, the grant to Canterbury from the Education Department stood at £8,645, of which £5,000 was for Arts and General, and £3,645 for the School. The following year the grant was reduced by £3,500 but it was the Arts and General courses which had to manage with less; the School of Engineering grant remained at the £3,645 level. Student fees, another source of income, were drastically affected. Fewer students could afford to attend a University and engineering students were especially handicapped when the Government first reduced, and then withdrew, bursaries. The *Press* declared it was "impossible not to condemn" such a step, "even in the present desperate situation."[2] Of the students who boarded at Rolleston House in 1929, 44 out of the 82 were studying engineering. The School of Engineering contribution to the Board of Governors' annual report revealed the steady decline in the numbers of matriculated students studying either for the degree or the associateship qualification. 129 in 1924, the numbers had dropped to 75 by 1935. In 1938, they had risen again to 84 but the outbreak of war the following year was to send them initially plummeting once again.

One feature of the engineering degree from its earliest days was the requirement that a student complete 18 months' workshop practice. The depression revealed that this whole matter needed to be handled very delicately. Even a scrupulous firm could dismiss experienced men and employ students in their stead, needing to pay them considerably reduced wages. There had already, in 1924, been complaints from the Amalgamated Engineering Union about such behaviour. By 1927, the Engineering Union was disturbed at the precariousness of its members, feeling strongly that College students were usurping their positions. A letter on behalf of the members of the Union was forwarded to the Minister of Labour stating: "Previous correspondence and data placed before you in connexion with the employment of students proves (sic) conclusively that even when trade is at its best, the question is a serious one to us. When, during a period of depression such as exists at present, our members are to be further victimised by the employment of people who, in a large number of cases, can afford to be idle for even a year or years, we are of opinion that sincere efforts should be made to prevent their employment until such time as those who follow the calling for a living are again fully employed . . . "[3] This letter revealed that practising engineers on the shop or factory floor had always felt a modicum of envy towards those students employed there during their vacations. The *College Review*,* facetiously advised aspiring graduates how to behave during their time at the Addington Workshops. A chance to make a student look ridiculous was, no doubt, on occasions greatly welcomed. However, it does not appear that graduates of this country were ever advised, as their counterparts in England were, to conceal from potential employers the fact that they had a degree. While there was undoubtedly a measure of ill-feeling between the engineering workers who were anxious not to have their jobs placed in jeopardy and the students, the whole issue was dealt with amicably. After all, it was irrefutable that the students were required to gain their 18 months' workshop practice and, on the whole, there was an excellent rapport between local industry and the School. While it was not possible to alter the terms of a Union Award, which protected the interests of the workers, a compromise was reached with local engineering firms around Christchurch so that good-will was restored.

At this stage of the history of the School, non-matriculated students still greatly outnumbered those matriculated students studying either for one of the Engineering degrees or for the Canterbury College associateship, although this was phased out in 1929. The non-matriculated, the "extra" students, were greatly affected by the financial hardship of the times in that many of the firms around Christchurch were forced to cease their system of sponsoring students at the evening classes. Their numbers declined from 149 in 1925 to 82 in 1934. Powell pleaded, in vain, that the Board would accept a reduction of charges in order to retain as many of the evening students as possible. "At one stage of its progress, the School of Engineering was practically kept alive by the Evening students," he wrote. He greatly deplored "the gradual falling off in the attendance which is partially due to the high charges.[4]"

Other economies were effected by converting the various Engineering Exhibitions into honorary awards which carried no financial assistance. This applied not only to the indigenous awards but also to the Exhibitions held by graduates with the United Kingdom firms of General Electric Company and British Thomson Houston. (These exhibitions had been established in 1919 and 1926 respectively). In this way, by a policy of cutting back here and pruning there, the School managed to carry on. With a greatly decreased income, the School was in no position to authorise any but the most basic repairs and maintenance to the buildings.

*see p. 56

Repeated requests that the School's office should be brightened were greeted with the reply that there was no money available for alterations. However, the subsidence of the hydraulics laboratory had to be treated seriously. Steele described the whole episode in a very low key manner: "When viewed from the outside the east wall appears to have bent outwards near the middle of its length", which he found "rather disquieting."[5] The matter was referred to Collins, the College architect.

If the buildings were barely maintained, equipment and apparatus suffered the same fate. The estimates provided for the School each year allowed only the barest running costs and the most basic materials; no new equipment could be purchased and ingenuity had to be exercised to keep what was there in reasonable working order. Powell's requests for interphones and a duplicating machine not surprisingly had been turned down. The 1933 annual report stated that there had been "no new apparatus . . . during the year . . . "no new books of reference had been purchased for some time."[6] When Robinson asked permission to buy a new theodolite he was informed: "the Board desires the utmost economy . . . all new developments . . . must stand over just at present."[7] Existing equipment had to be maintained; the 50 ton Buckton testing machine purchased in 1895 required repairs and adjustments. And as Steele indicated, "there [were] a number of machines in the laboratory over 30 years old which are beginning to give trouble."[8] Whichever way it looked, the Board could foresee major expenses if standards were to be upheld.

For a while staff was maintained at the pre-depression level, but eventually a policy of attrition set in. R.A.McLennan had held a position renewable annually in the Electrical Engineering Department calibrating instruments. Eventually in 1930 he had to be dispensed with. D.W.D.Hutcheson from Mechanical Engineering likewise had to be dismissed with regret in 1932 and there were no allowances for student assistants. The existing staff was drastically overworked, nor was this just confined to the range of their normal duties. Despite earlier requests by Scott that the Board should employ a licensed surveyor when any of the endowment lands needed mapping, which at the time was acceded to, during these depression years, Robinson was once more required to undertake the surveying of a block near Ashburton. His failure to comply as promptly as the Board desired called forth a sharp reminder. All of the staff had to submit to a considerable decrease in salary as part of the nationwide economic stringency, as well as contribute a certain percentage (initially 3d in every 1/8d) towards the Unemployment levy. The drop in living standard was considerable and obviously caused especial hardship for the less established and more junior members of staff. At one point Steele put in hopefully for a raise for Rastrick, on the grounds that he was getting married at the beginning of the following year, but to no avail. It was not that the Board was callous — as landlord and employer of considerable numbers of people, it was very aware of the widespread hardship that was being endured — but it had no option. At least Rastrick and the others had some guarantee of tenure, even if they had to perform what was a form of sweated labour for their monthly salary. When the fireman at the School retired, there were 237 applications for his position listed with names and addresses.

One staff member, however, did get short shrift and that was the School of Engineering typist, Miss McIntyre; she had been with the staff for a number of years after the amount of clerical work from the School had become too much for that dogsbody Morrison and a boy clerk. Scott had requested a male clerk, but had been forced to settle for a "typiste", as it was called in those days, at half the salary and Miss McIntyre was installed. Attempts to subsume the work of the typist into the general duties of the Rector's office staff were always strenuously resisted on the grounds that it was highly skilled work, that many of the engineering records, eg, those relating to the practical work the students had to do as part of their degree, had to be on the premises for easy referral and so on. However, come the depression and the need to retrench, and it was decided that the only way Miss McIntyre's services could be retained was if she would accept a 50 per cent reduction in salary. This brought her salary down from £130 to £65 a year. She refused, resigned, and her resignation was accepted "with regret."[9] The Board shed crocodile tears and replaced her with a temporary typist at a salary of £52, ie £1 a week. The first one proved unsatisfactory but the second, Miss Gundy, was eventually promoted to the Rector's office where she was secretary for the Professorial Board.

Thus, the School carried on against difficulties but the real legacy of the depression years was the entrenched conservatism generated by the years of financial stringency. Instead of adventurousness, an eager going out to meet new ideas, creativity and innovation, caution and self-protectiveness reigned. Not surprisingly nobody was willing to take risks or to experiment, and the School and the country at large suffered accordingly. It is a sad commentary on human nature that it took the second World War with all its wastage of life and property to push the scientists and engineers, not just of New Zealand, but of the world, into greater technological advances.

As the depression lifted, it looked as if the School of Engineering could begin to function fully again. But inevitably the depression years created a backwash in that as fewer students graduated, there were unfilled engineering positions within the community once the economy had stabilised. In the annual report of 1936, Powell, the current Dean of the Faculty, reported: "A striking feature of the year has been the large number of enquiries from persons and firms both locally and abroad desiring to employ Engineering graduates or students. The demand has been greater than the supply and it is likely that there will be a shortage of young qualified engineers during the next few years. The comparative smallness of the enrolments of new students for the earliest year of the course will accentuate this shortage."[10] The same disproportion between supply and demand was the tenor of Powell's reports to the Council over the next few years.

CHAPTER SIX

Provincial Rivalry

If the University of New Zealand could be described as "the child of discord," then the "ancestry [of the so-called 'special schools'] was but too evident. They were cradled in controversy; they existed on jealousy and government grants. In no way else is the colonial, the provincial origin of the University so manifest, in no way else has it borne such exasperating fruit."[1] Provincial rivalry was a strong motive behind the original decision to found a School of Engineering, and it did not just disappear once the School was firmly established. If one of the most telling arguments in favour of setting up the School in 1887 was that otherwise Otago would snaffle the coveted prize and chalk up yet another special school to its name, it became, as the years passed, equally pressing to prevent Auckland University College from sharing the prestige, importance, and the financial benefit of an Engineering School. It became a very emotional issue, and at one point even produced an unpleasant, if facetious exchange of letters between the Registrars of the two Colleges, Auckland complaining that Canterbury's advertising of its Engineering School in the Auckland newspapers was an unwarranted intrusion, and warning Canterbury "to keep off the grass."[2]

This is not the place to go into detail about the origin of the Auckland Engineering School, its emergence from a derelict School of Mines, nor its continuing battle with insufficient staff and poor accommodation. Indeed there is something admirable in the tenacity with which it hung on, even against most unlikely odds. Canterbury, however, always saw its role as that of protecting the interests of the engineering industry at large and of the country as a whole. It was able to absorb great quantities of Government money, at the same time declaring piously that to establish a second school would be an unjustified extravagance.

As was to be expected Professor Scott threw himself gleefully into the fray. Sadly he must be convicted of the ability to argue persuasively from both sides of any argument as, when it suited him, he could roundly declare how much the country was in need of qualified engineers but was not above stating that the country already had as many as it could deal with, and that the one or two graduates which Auckland could turn out in a year would flood the market. Scott for 20 years represented the Canterbury College Board of Governors on the Senate of the University of New Zealand where he had ample opportunity as convenor of various committees to do with science, engineering and mining to block the presumptuous claims of Auckland to have its School of Engineering recognised. This is where the bizarre situation of overseas examinations assisted Canterbury in its efforts to keep Auckland at bay, and it is noticeable that Scott was one of the last of the Canterbury professors to vote in favour of the discontinuation of the system of overseas examining. The constituent colleges were responsible merely for teaching and had no hand in examining their students; the power of granting or withholding degrees was invested in the Senate of the University of New Zealand, a body which was dominated in its early years by southern representatives who saw no reason for encouraging Auckland's pretensions. The recognition obtained by Scott for the Canterbury degrees by the prestigious United Kingdom Institution of Civil Engineers was also very much a trump card. He was able to make much of the fact that it was the Canterbury School which had been thus accepted, not the "bastard institution"[3] at Auckland. It was many years before the Institution accepted the Auckland degrees. This was a cause of considerable acrimony, Auckland claiming that Canterbury had deliberately excluded them, and Canterbury smugly repudiating such a claim, without completely denying it.

However, the fact that Auckland's persistence in the end won out can be partly placed at Scott's door. In 1913 there was a strong move afoot to establish a School of Architecture in New Zealand and the logical place was obviously Canterbury, although there were other strong contenders for the honour, notably Victoria. Much of the political manoeuvering went on while Scott was overseas on his third trip to Great Britain. When the need to establish a School of Architecture in the Dominion was first mooted, S.Hurst Seagar, a notable Christchurch architect, at that time lecturing in architecture at the School of Art, was very prominent in pushing for the School to be at Canterbury.

He declared that it would initially require very little outlay of money, as so many of the necessary courses doubled with those offered by the Civil Engineering Department. Ultimately it would only be a question of appointing a Professor of Architecture. On his return, however, Scott was not enthusiastic, and in the end he allowed, even encouraged the School of Architecture to go to Auckland with the condescending implication that if it had that, then perhaps it would stop yapping for the larger reward of a School of Engineering. On the surface this ceding Architecture to Auckland looked a shrewd political move, even altruistic. It is hard, all the same, not to query Scott's motives, remembering how fiercely he fought to prevent the establishment of a Chair of Physics and any encroachment on his territory. It certainly seems possible that he quite frankly did not want, within "his" School, a Professor of Architecture who would have a measure of autonomy. The easiest way out of the dilemma was to grant Auckland the School of Architecture. If this is the case, he had failed to foresee one major difficulty. The prescriptions for the Architecture course — and as convenor of the Senate committee which determined them, he was fully responsible — laid down in black and white in the University calendar: "the same as for Civil Engineering."

Auckland was not slow to realise the advantage it had gained and inquired why the Senate was willing to grant degrees in architecture but would deny an engineering degree to a student who sat and passed exactly the same examination if that examination were labelled civil engineering. This argument was unanswerable but was thrown away by the introduction of a great number of emotive side issues. If Auckland students were disadvantaged by having to attend Canterbury College and incur the expense of boarding, so after all, were students from the rest of the country. The Auckland detractors were not slow to point out that a School at Auckland would benefit only students from the city of Auckland, while those from the province of Auckland would still have to board. They suggested that to provide travelling and boarding bursaries would cost the country far less than equipping and building a second School of Engineering. The other emotive argument could equally well be demolished. Auckland set great store on the energetic developments of the engineering industry within its province, and its assumption of superiority did not go down very well. This is not to suggest that there was any lack of special pleading on the Canterbury side. The local papers, while deprecating Auckland's pretensions, were most sanctimonious in the pride they expressed that *they* were not tainted with any tawdry provincial competitiveness. However, if Auckland had stuck to the main issue which involved a point of justice, instead of clouding matters with provincial protectiveness and superiority, it might perhaps have won its battle for recognition sooner. As it was, little by little Canterbury's unassailable position was undermined, although in the end it was not until 1951 that Auckland gained full recognition for all three branches of the engineering degree. This victory for Auckland was seen as a defeat for Canterbury. (An amusing insight into the strength of the rivalry between the two schools is provided by a statement made by Auckland's Professor T.D.J.Leech, that he had "twice turned down the offer of a chair at Canterbury."[4] These offers were allegedly made in 1941, and 1943. Certainly the chair of Civil Engineering was vacant in 1941, although it was decided to defer making an appointment until after the war because the range of applicants was not satisfactory. Leech is not recorded as having applied. Nothing more was done until the end of 1945. His declined "offers" must therefore refer to personal comments from some staff member or other. They were not official, but they have become part of Auckland's "myth.") The Canterbury School of Engineering, after enjoying so many years of supremacy, was forced to watch helplessly while the Auckland School forged ahead. Past reputation was to prove insufficient in the increasingly competitive world which emerged after the Second World War and as the Canterbury School deteriorated, its demands for consideration sounded increasingly querulous. The ultimate outrage was the Government decision in 1952 to assume that the two Schools should produce equal numbers of graduates.*

*see p. 208

Canterbury University College in 1940. The School occupied the buildings in centre foreground. See Appendices for detailed floor plans of the School on the city and Ilam sites.

CHAPTER SEVEN

Canta

Dissatisfaction at the quality of teaching offered by the School made its first public appearance in an anonymous article in *Canta* in October 1941, under the striking heading: WHAT'S WRONG WITH THE ENGINEERING SCHOOL? Sheltering behind the pseudonym, "Engineering Graduate", the writer praised the "organising ability, perseverance and drive of the late Professor R.J.Scott" and deplored the "distinct slowing down of progress" since his retirement. So much so, indeed, that "there has been a tendency to live in the 'glory that was Greece and the grandeur that was Rome.'" The article focussed on the inefficiency and uninspiring nature of the courses, stigmatising the overcrowding, the obsolete equipment, and the lack of any real verve or initiative. Gauging accurately those areas which were most vulnerable to attack, the writer warned that "for many years now, under able and energetic direction the rival School of Engineering at Auckland University College has been forging ahead . . . Even now, A.U.C. School of Engineering is engaged on important research for defence work." The absence of research, the entire absence of workshop facilities, the obsolete machines as well as diagrams and slides, "with dates such as 1886, 1901, etc, . . . " were attributed to "the 'laissez-faire' policy which has prevailed in recent years". The Heat Engines laboratory with "its collection of derelict and obsolete machines" deserved rather the title of Heat Engines Museum. Against this, the writer did concede that "the Hydraulics laboratory stands alone as a well-equipped laboratory, and with its classes well organised."

"Engineering Graduate" was scathing about the policy of rotating the deanship every two to three years, while most of the remainder of his article dealt with the poorly lit buildings, the depressing classrooms and the general run-down appearance of the School. "To sum up what is wanted is a vigorous and go-ahead policy which would organise research, bring laboratory equipment up-to-date, establish a workshop for students, and also bring lectures and diagrams up to the minute. A 'what was good enough for grandpa' policy has no place in the modern world, when engineering like all other branches of technology is marching ahead."[1]

This attack on the School of Engineering by a former student attracted considerable attention. The *Press* was quick to fasten on the reference to Auckland's School of Engineering: "in its alleged backwardness the Canterbury College School of Engineering provides arguments for granting the Auckland University College School of Engineering the right to take students for the full degree course. This could have serious repercussions for Canterbury College, as more than half of the present engineering students come from the North Island."[2] Powell, as Dean of the Faculty, was asked to comment on the *Canta* article: "I can agree with no more than the general principle that some things could be improved in the School of Engineering. Finance is the difficulty. We are more or less living from hand to mouth. The article may serve a useful purpose in reminding people that there is a School of Engineering and that if Christchurch wants to retain it the people must take a bit more interest in it."[3]

"Engineering Graduate" was only saying in a more succinct, and public form what had already been said in different ways and at different times since Scott's retirement. When the complaints had issued from the Dean or one other of the engineering professors, accompanied inevitably by requests, explicit or otherwise, for money, the Council had found it possible, even expedient, to ignore them. Now that a recent graduate had ventured to throw down the gauntlet, it became necessary for the Council not only to pay attention, but to be seen to do so. But, for all the frenzied activity which followed this article in the form of a plethora of sub-committees, it was not until the following year that a rebuttal by the Rector appeared in *Canta*, in an edition labelled EXTRA. Inset into Dr H.G.Denham's* "Answer to Criticism of Engineering School" was an apology from "Engineering Graduate" for any offence he might have given in his references to the present members

*Denham, Henry George, 1880-1943; b ChCh; educ CBHS, CC (MA, 1905; MSc, 1906), awarded 1851 Exhibition Scholarship, studied at Liverpool (DSc) and Heidelberg (PhD); lect at Queensland Univ, 1912; Prof Cape Town Univ, 1921; Professor of Chemistry, CC, 1923-43; Rector, 1941-3; hon m Soc of Chem Industry; m Ac Bd and Sen UNZ; m and cm CACBG.

of the staff. (It is rather strange that, despite the lapse of a year, Denham should refer to the "recent" article. It is also the case that Denham quoted an alleged commendation about the School which does not occur in the original article; thus Denham claimed that "the writer of the recent criticism is himself subconsciously aware of the aims of a University education when he states that 'it is generally agreed within the engineering profession in the Dominion that the C.U.C. School of Engineering produces a graduate with a substantial background of technical fundaments (sic) in the main categories, Electrical, Mechanical and Civil. Admittedly the average graduate finds it possible to deal with practically every sphere of engineering after finding his feet.' This is indeed, high praise, and shows that in spite of any shortcomings the School and its Staff has (sic) delivered the goods." Unfortunately these words did not appear in the original article. Either there is a second article, containing these words, which has unaccountably gone astray or we can deduce that the Council was rattled.) Denham followed this by declaring: "it should not be assumed that, because the Council has not advertised its plans and projects nothing is being, or has been done. Numerous conferences have been held, a good deal of spade work accomplished, and when the vexed question has been settled as to whether the resources of New Zealand are sufficiently large to justify running two fully equipped schools of Engineering when one of them is capable of turning out nearly three times the average output of the past twelve years, the worth-while changes so forcefully advocated by 'Engineering Graduate' will automatically fall into line."[4] This may sound like obfuscation but in fact it was not. Until such time as the Government would guarantee capital grants to the School of Engineering for upgrading of buildings and equipment and increasing the staff levels, there was very little that the Council, or the School could do to improve conditions. And inevitably there was no finance available while the war lasted. From then on, until the move to Ilam, there would be a policy of makeshift and make-do, and the School would have to exercise considerable ingenuity and its staff be stretched to the uttermost, while everything hung fire.

Meanwhile the Council, as part of the "spade work" which Denham had mentioned, had written not only to the three professors, Powell, Steele and Calvert, but also to Pollard,* who had been seconded to

---

*Pollard, Thomas Royle, 1899-1969; b ChCh; educ Marist Bros School, CC; jr 1 CC 1922-41; sl 1941-48. Director C.C.I.D.D. 1946-65; founder and head of Consumers' Institute Laboratory 1965-69.

the D.S.I.R. at the outbreak of war. As Director of the Radio Development Branch, he had considerable responsibility, and much of his work was classified top secret. His energy and expertise had deservedly won the respect of the Council. At the time of the *Canta* article, he was well into his stride with his Radio Defence work. When Denham wrote in his rebuttal of the *Canta* article: "there is no question that the workshop of the School, under the direction of a member of the Staff made available by the Council has done work of the greatest national importance,"[5] he was referring to Pollard. Accordingly, when the three existing professors were asked to comment on the validity of "Engineering Graduate's" derogatory remarks, Pollard was also consulted. All four were asked to outline their proposals for the future expansion and development of the School. Pollard went further; he recommended that recent graduates should be canvassed for their opinions on the current state of the School. The Council adopted this suggestion; each of the four submitted the names of two recent graduates who were then asked to give their confidential opinions of the four proposals. One of the most vociferous replies received was from W.W.Young, one of the two men nominated by Pollard. Young was closely connected with Pollard's war work. It has been suggested on good authority that Young was the anonymous "Engineering Graduate", which leads us to the inescapable conclusion that possibly he had been nudged into writing the attack by Pollard himself.

The Council must have felt that it had been handed a hot potato when the majority of the comments from recent graduates arrived. "I consider that with the retirement of a greater portion of the present School of Engineering staff and its substitution by a carefully selected group of forceful, energetic and competent men many of the present troubles connected with the School would disappear"[6] was the expressed opinion of N.B.Manssen. Others agreed that members of the present staff, especially Powell and Steele, were "perfect gentlemen",[7] but no longer adequate — if they ever had been, judging from L.B.Wood's strictures in the past — to offer leadership in the greatly expanding field of engineering education. W.W.Young stated in the conclusion of his long analysis of the problems besetting the School:

> In case my criticism should appear to convey personal attack in any way, I wish to make it clear that at no time have my relationships with the School staff been other than very pleasant. In the interests

of Canterbury College itself, I consider that the present problem should be handled quite independently of the destiny of the existing staff, for it is far more important to take care of the future than to try to perpetuate a brilliant but rapidly fading past.

His final phrase echoed the sentiments of the anonymous *Canta* article. But while his conclusion was couched in tolerant and non-judgemental terms, the main body of his detailed criticism of the School contained astringent comments about Powell and Steele. Powell, for instance, "does not adopt any form of progressive policy, in spite of continued requests on the part of students and graduates." As for Steele:

> I consider that the professor is absolutely unconscious of what is going on in the engineering world outside, even in Christchurch. His associations with commercial engineering must be practically nil. His inability to discuss engineering matters in an intelligent fashion would lead one to believe that his knowledge and experience are negligible. His personality is decidedly weak, and his outlook remarkably limited.[8]

Powell's and Steele's written responses to the *Canta* article were consonant with the kinds of criticism levelled at them. Powell claimed that "the School of Engineering and the course in Engineering is not perfect, but I have always considered that improvements should be introduced gradually."[9] Steele, on the other hand, gave the impression that quite frankly he did not understand the real cause for dissatisfaction. In today's terms, it was as if he had been told that the School was at a disadvantage because it had not installed the most up-to-date computers, and he had protested that, although old, the typewriters were completely adequate. There was a real sense that he had been left behind. But then this was the man who, when asked about the extent of research in the Mechanical Engineering Department, replied a little condescendingly that he thought "all the research needed in mechanical engineering had been done."[10] Even the accounts of his stocktaking in his department reveal him to be humourless and pedestrian: "One steam engine, one torpedo boat engine,* one pair of pliers, handle broken."[11]

*This was just one of the many pieces of equipment which Scott had been able to requisition for the use of the School. In this case it was Government surplus. He had hoped to acquire the whole boat but before he could take possession in 1907, it went aground off the coast of Dunedin and he had to be content with the engine. Its present home is the entrance hall of the School at Ilam.

The Mechanical Engineering Department certainly offered confused and outmoded instruction, but it would be most unjust to lay the whole burden of responsibility on Steele's shoulders. At the risk of wearing out the same refrain, it would be well to recapitulate some of the difficulties faced by that department which were built-in features from the outset. Scott's bias towards mechanical engineering meant that it was the first engineering laboratory established at Canterbury College. Reminiscing about his student days, Professor Rastrick described the state of engineering when he first enrolled in 1923 as "mechanical engineering with the other branches, electrical and civil, grafted on to it."[12] Consequently, the greatest burden of teaching, whether of matriculated students or Scott's "extra" students, fell to the Mechanical Engineering Department. No wonder a man as qualified as Dr Grundy should complain at the standard of teaching expected of him. First established in the last century with the much-publicised steam engine and testing machine, the Mechanical Laboratory inevitably was the last to be updated. Indeed, it suffered greatly from the moratorium on imports imposed by the difficult years of the depression and the war. Likewise the testing programme, undertaken initially with such pride and enthusiasm by Scott, became an increasing burden as the years passed. If Steele could express incredulity about the existence of research in mechanical engineering, this can partly be attributed to the dread load of commercial and Government testing. Overseas visitors expressed their astonishment that a University should undertake such testing, but it was a remnant of New Zealand's colonial days before industrial firms had expanded sufficiently so that they could set up their own testing procedures. Apparently, Universities in Great Britain or the United States accepted only such testing as worked in with their own research projects. However, Steele had joined the staff straight after graduation, and had never encountered any large engineering enterprise. Apart from attending a conference in Adelaide in 1936 he had never left New Zealand. On the other hand, the accusation in the *Canta* article that, unlike Auckland, Canterbury was not contributing to the war effort, was most unjust. The Mechanical Engineering Department had to endure an increased load of routine and essential testing for the Ministry of Munitions during the war, none of which was dramatic or eye-catching.

While the stagnation of this department was serious in itself, it was doubly serious when so much of the teaching in the early years of the engineering degree relied on mechanical engineering. In this way, the other two departments carried a similar weight of inertia, although civil engineering was less affected. Electrical engineering suffered much the same fate as mechanical: the weight of the Scott tradition, the lack of money for new developments or equipment, and the responsibility of the evening classes. Other problems, according to W.W.Young, were the inefficiency of presentation in the laboratory teaching for Electrical Engineering I and II: "nothing short of a bedlam," and involving "long hours of work on laboratory books in other than laboratory time."[13] Such changes which had taken place consisted more of a re-arrangement of existing courses than the introduction of anything dramatically new, (until, that is, Pollard's course in electronic engineering.) In 1930, the School's annual report referred to "the new courses in Electrical Engineering providing for two alternative degrees, (a) Electrical (Mechanical) and (b) Electrical (Civil)." These had finally gone through the slow University procedure and been "approved by the Academic Board and Senate."[14] It is hard to imagine exactly what advantages this bifurcation of the Electrical degree conferred. The recent graduates consulted at the end of 1942 certainly felt it should be dispensed with.

Another major cause for concern was the policy of rotating the Deanship every two to three years. In this way, Powell and Steele had a heavy administrative burden added to their normal academic duties. It is not surprising that the tone of Powell's letters should become increasingly petulant. As Steele approached his retirement age, he declined accepting the Deanship so that Powell had to endure a further spell as Dean. During the time that Cull was Head of the Civil Engineering Department, there was no move to appoint him Dean of the Faculty. This omission gave rise to considerable speculation. Cull's suitability for the role was questioned, especially as one of the responsibilities connected with the Deanship was to attend the Senate's Academic Board of Studies. There, the Canterbury representative was expected to play a strong part in warding off Auckland's moves for its own Engineering School. It was therefore assumed that Cull could not be entrusted with such a responsible task. However, as it turned out, the reason why he was never Dean is perfectly simple. He always flatly refused the position, maintaining that he needed to give all his energies to the Civil Engineering Department. Similarly he declined a nomination as the Canterbury representative of the newly formed New Zealand Standards Institution in 1936, offering the same reason: "with the present staffing of his department he cannot give the requisite time to the Institute's work."[15] Thus, it was not until Calvert was appointed in 1945 that Powell was released from the treadmill. If Steele and Powell were able to bring some enthusiasm and interest to the position in the first years after Scott's retirement, it is hardly surprising if they had lapsed into lethargy after repeated spells as Dean of the Faculty.

None of the submitted reports following the *Canta* article contained any major criticism of Calvert, who, at this time, was in the invidious position of Acting-Professor because of Cull's retirement and the delay in appointing a successor. It does not appear, however, that this was the sole reason for restraint. On the whole, the Civil Engineering Department fared much better than the other two, even earning positive commendation for its attempts to modernise teaching methods and keep abreast of new developments in the theory and practice of civil engineering. Overseas examiners' reports frequently commented on the thoroughness of the teaching of hydraulics.

The *Canta* article and its aftermath produced not only a great deal of consultation among practising engineers throughout the country, recent graduates, members of the Institution of Engineers and sub-committees of the College. As well, the Council felt it was advisable to bring in an expert in the field from outside New Zealand. Accordingly, it invited Sir John Madsen, Director of Australian Scientific Research Liaison to inspect the School of Engineering. Sir John visited in 1943. In his submissions to the Chairman of the Council, he stated vigorously that "the closest co-operation should be maintained between the pure and applied section of scientific work."

> Your Engineering School possesses the nucleus of equipment in its respective laboratories which is of quite good standard. The time, however, has arrived, at which this nucleus should be expanded so as to provide for developments, particularly in the application of Physics and of Chemistry in the modern industrial field. To meet this need . . . your College should seriously consider these developments, and I would recommend that the following steps should be taken:
>
> (a) The establishment in Engineering of a lectureship in Communication Engineering, together with the necessary laboratory, equipment and maintenance.
>
> (b) The taking of preliminary steps to provide for the future development of Chemical Engineering.

The remainder of his report covered a wide range of topics: Administrative Control, Research, Practical Training and so on. He doubted whether the 15-day drawing examination was of any significant value, "whether such an ordeal [was] really necessary."[16] But obviously the crux of his recommendation was that Chemical Engineering and Electronics should be established with as little delay as possible.

The School of Engineering might be "unenterprising,"[17] and given over to "mass production"[18] methods of educating students, but while the war lasted, little could be achieved. At the outbreak of the war, the numbers of students attending the School, degree students and "extras", had fallen significantly as men were called into active service. As the years of the war dragged on, serious concern was expressed throughout the country that branches of study such as medicine, dentistry, mining and engineering, which were crucial to the country's well-being and economic stability, were being denuded of students. The Senate of the University of New Zealand made a representation to the Government in 1942 that a small number of student places should be allocated each year in these various important specialisms so that at the end of the war there would still be a continuing and suitably educated manpower. The Government accepted the necessity for this. (Further representation, initiated at Otago University, that it was disadvantageous to ignore the significance of especially brilliant students in the Faculties of Arts, Law, and Commerce fell on deaf ears. Indeed, the three other University Colleges did not support Otago in making out a special case for the humanities.) So, after 1942, engineering students could apply for deferment of their military service, provided they maintained a steady progress in their studies. This kept the numbers of matriculated students at a high level, but non-matriculated apprentices who attended evening classes at the School of Engineering were not eligible for this exemption. The numbers of these students dropped dramatically and indeed, the initial roll at the beginning of a year, eg, in 1942, 73 out of a total 170 enrolments, gave a false indication of the numbers of students completing the year. Apart from those called up during the year, unless their employers were able to make out a strong enough case for them to remain in essential positions, many students were unable to continue their courses because of the demands of overtime. The drop in student numbers had the inevitable economic spin-off in the reduction of student fees.

Of those members of the staff who were required for military duties, the absence of T.Stanton, appointed lecturer in civil engineering in 1939, was the most disastrous. The Army required him to supervise fortifications for the protection of Lyttelton Harbour against possible enemy attack and rejected the College's appeal. The utmost concession it would grant was to release him for teaching one day a week. His absence meant that the Civil Engineering Department was placed in a very precarious position and Calvert was shrill in his alarm. Fortunately, for one year, R.A.Campbell was able to return and give lectures on the Principles of Civil Engineering and in 1943, J.K.Martin was appointed in a temporary capacity.

As for the other departments: Mechanical Engineering bore the brunt of the greatly increased testing, especially for the Ministry of Munitions, but had no direct problem with staff being co-opted into government work, despite an abortive attempt by Professor T.D.J.Leech, in his dual capacity as Dean of the Faculty of Engineering at Auckland and Deputy Director of Scientific Developments (Engineering) for the D.S.I.R., to second Lush and Bamford. Electrical engineering, early in the war, lost the services of Pollard who became Radio Engineer for the special research section of the D.S.I.R. Radar research was the most direct contribution to the war effort by Canterbury College, as, even though Pollard was paid by the Government and answerable to the Government, his researches were undertaken on College grounds and encroached considerably on College property. All the engineering departments were greatly inconvenienced by the D.S.I.R. taking over the College workshop and a fair percentage of the working time of the engineering technician H.B.Wrigglesworth. When he was called up, the appeal lodged by the College on his behalf was successful.

Even if there had been money for more staff, both academic and technical, there were no men available for the positions; they had all been either called up or diverted to essential services. The School had to struggle on as best it could. And one of the ways in which, over the years, the School had dealt with the problems of new equipment was for it to be designed and made within the Engineering workshop. The war-time staff, however, scarcely had time to maintain existing machinery in good working order, let alone construct new apparatus. And it was totally impossible to import from overseas because, even if priority passages could be secured, the major engineering firms in the United Kingdom were all engaged in manufacturing armaments and munition. It was a stalemate.

The end of the war brought more difficulties and hazards as a result of the influx of returned servicemen which had somehow to be accommodated. They brought their own problems in that they fell into three categories: men with no engineering experience prior to the war who could scarcely be treated like the youngsters entering the School straight from high school; those men who had partially completed their courses and wanted to resume them; and those who felt they needed a brief spell of refresher study before taking up positions in industry. The difficulties of overcrowding, of duplicating and reduplicating courses were compounded. The initial criticism from an engineering graduate in *Canta* had spread to all sections of the engineering community. It was fully recognised that until the School could institute new courses, encourage research and acquire new equipment it could not hope to attain its earlier privileged position, but nothing could be done without firm Government assurance that finance was available. The School had, in effect, reached an impasse. In April 1946, J.H.E.Schroder, Chairman of the Council, wrote to the Commissioner of Works, E.A.McKillop, about the current status of the School of Engineering:

> . . . you may be inclined to agree with me that the past matters less than the present and the future. The central fact is that the Council has for a considerable time been thinking hard about reconstructing the School, regards this as the crucial period — within the next 12 months all three Chairs will have been newly filled — and is especially anxious to do two things, which are closely connected. They are, to bring the School into close and continuous touch with the public and private engineering work and leadership of the country, and to develop the research and experimental side of its function. In fact, some advances have been made; eg, in establishing Pollard's electronics branch, in the chemical engineering line, and in encouraging Calvert's hydraulics experimental work.[19]

The exciting new courses in Chemical Engineering and Electronics which were to be introduced in the next years, and the continued expansion of Civil Engineering were thus the prelude to more fundamental re-structuring which could only take place once the School had taken up its new quarters at Ilam.

CHAPTER EIGHT

Civil Engineering

The experimental work undertaken by the hydraulics laboratory drew commendations on all sides. Such well-deserved praise was significant, revealing that even if the other two departments had stagnated in the middle years of the School's history, the Civil Engineering Department had shown itself to be dynamic and resourceful. Considering the appalling conditions against which the School was struggling, Civil Engineering's achievements become even more remarkable. And Calvert, in the years of his "caretakering" of the department, was merely carrying on the tradition which Cull had established, although he was more fortunate than his predecessor in receiving the Council's support for his endeavours.

Cull was appointed to the chair in 1930 and almost immediately he was co-opted as chairman of a Government enquiry into the Napier earthquake. On his return he lost no time in reminding the Council that the Civil Engineering Department required more staff if it were to keep pace with the rapid changes taking place in civil engineering science and practice. He regarded his department as drastically understaffed, and therefore believed he must devote to it all his time. (Certainly it was the case that the Civil Engineering Department attracted a much greater proportion of the students, even after the division of the B.E.(Electrical) into Electrical-Mechanical and Electrical-Civil.) As a result he refused, to the obvious annoyance of Powell and Steele, to assume his share of the administrative responsibility of the School. Perhaps because of this single-mindedness, he always appeared more forceful and forward-looking than his two colleagues in his written communications with the Council. As well, he did not have to carry so great a weight of the Scott tradition. He certainly deserves commendation for his efforts to develop civil engineering during the years when money was scant and progress at a standstill. He wrote letter after letter asking the Council to provide more money so that he could appoint more staff and introduce vital new developments. By contrast, there is a decided lack of such requests from the other two departments. But for all his clamouring, the only concession he could win from the Council was reluctant permission for R.L.James to give a series of six lectures and demonstrations on soil properties in 1937. (As was later to be the case with Alley's soil researches, the Council had only accepted that James undertake his research when it was quite clear that there would be no financial drain on the College's scant resources.) James's lectures were well attended, with "an average attendance of 37 including engineers from the city. A gratifying feature has been the attendance of engineers from the Railways and Public Works Departments, the City Council, Drainage Board and Gas Works."[1] However, Cull was still protesting vehemently that it was becoming "less and less possible to do any original work in the department owing to the shortness of staff."[2] In 1939 he once again made a strenuous plea for "strengthening the staff of the Civil Engineering Department."

> With the present establishment, the Civil Department is just carrying on but I cannot say we are progressing . . . We should be playing a greater part in advancing the standard of Civil Engineering practice in this country and co-ordinating our work with that of the great engineering departments of the state.
>
> In all we do we should feel we are working towards some goal. During nine years of my tenure of office, modifications have been steadily going on in the courses of instruction, amounting in the aggregate to a considerable change, and I have endeavoured to keep every alteration in line with a definite policy of progress. Further important changes in the course are now pending, and I would be much strengthened by some expression of the Council's policy of development.[3]

Despite all his appeals, the Council was not forthcoming. Cull retired in 1941. In 1942, and then again in 1944, Calvert, as Acting-Professor, was quoting this identical letter in support of his request for additional staff "to cope with the increased teaching laid down for 1945."[4] (Surveying had become compulsory for all students, and there was the greatly increased engineering roll with the end of the war. In addition, a five year B.E.(Honours) degree had just been instituted. Later there had to be representations to the Government to ensure this degree was recognised in salary scales. If students with this degree received no more salary than if they had taken out an ordinary B.E. degree, there was no incentive for them to study for the extra year.) The frustration felt by Cull in 1939 had not dissipated; it had built up even more strongly. Without the immense support and co-operation of practising engineers, who gave either short courses of lectures on their particular speciality — it would appear that the Civil Engineering Department had more distinct specialisms at this stage, which made the loss of any one staff member much more disastrous — or who, like R.A.Campbell, the former professor, undertook full responsibility for Stanton's teaching in 1942, there were occasions when it might have been necessary to suspend the teaching of civil engineering altogether.

Despite difficulties with insufficient staff, when the Professorial Board, in 1943, instituted an enquiry into the research activities of the College, Calvert was able to provide an impressive list of his department's achievements:

> (a) Demonstration of applications of theories of aerodynamics in the study of wind pressure distribution on aerofoils and structures of different shapes . . .
> (b) Experiments on models for proposed hydraulic structures in connection with irrigation and water supply problems occurring in actual practice.
> (c) Experiments on models of rivers to study the model laws regarding problems of erosion and control of river banks, etc.
> (d) Experimental investigation of structural problems. A shaking table is now being constructed for research in problems of design to resist earthquakes.[5]

It is not surprising, when all this energy and enthusiasm were being directed into the department, that civil engineering continued to draw the students. In support of Calvert's application for the chair in 1945, Tocker commended him for maintaining "the standards and prestige of his department . . . It is very noticeable that Civil Engineering is much the most popular of the engineering courses among the students."[6] While this provides a good indication of the quality of teaching, at the same time it reflects the vitality of civil engineering throughout the country as a whole. A prominent feature of the Labour Government which came to power in 1935 was its decision to inject a large amount of capital into public works: roads, bridges, hydro-electric schemes, new aerodromes. Inevitably this brought engineering, and especially civil engineering to public attention. More students, attracted by good job prospects, came to study at the School. One unsatisfactory consequence of this upsurge in civil engineering throughout the country was that qualified engineers could command higher salaries with the P.W.D. than those offered by the College Council, which exacerbated the staffing difficulty which confronted the department. Although differential salaries of any kind were regarded with great suspicion, it eventually became necessary to advertise lectureships in civil engineering at £100-£200 higher than the going rate for the other departments. Even so, it was not until after the move out to Ilam that the department, and indeed the School as a whole was freed from the constant hassle of insufficient staff.

CHAPTER NINE

Chemical Engineering

Sir John Madsen, after his visit to the School of Engineering in 1943, recommended that the Council should introduce a course in Chemical Engineering at the College. He was, however, only endorsing the strongly held views of Professor H.G.Denham, the Head of the Chemistry Department. In an address to the Engineering Society of the College, as early as 1924, Denham had advocated "the training of Engineer Chemists as well as Chemical Engineers, the former being 'Chemists with a wide knowledge of physical Chemistry and imbued with the spirit and method of the best type of Engineering, and the latter being fully trained engineers — the more Chemistry they know the better it will be'."[1]

The depression was hardly the appropriate time for seeking Government support for a new venture, so it was not until October 1937, that Denham, with the full support of the Professorial Board, submitted a memorandum to the Council outlining the pressing need for Chemical Engineering in New Zealand:

> During the past twenty years tremendous strides have been made in the direction of providing men who have the equipment of the engineer but the outlook of the scientist. Experience has shown that industry requires trained men for its successful operation — men who are not merely technicians in the sense of the engineer but are also capable of understanding and correlating to the fullest extent the many processes they may be called upon to supervise. The chemist lacks the sense of proportion given by large scale operations, the engineer too frequently concerns himself rather with the construction of the apparatus than with the manner in which it operates . . . The suggestion is made that the Council explore the possibility of securing

the necessary finance to develop such a course, after convincing itself that there is a reasonable scope for the development of such a course in New Zealand.[2]

The Council responded favourably to the suggestion, deciding "to approach the Minister of Industries and Commerce with a view to ascertaining if the requisite financial assistance would be available, should such a course be developed here."[3] Denham was requested to sound out the opinion of the Secretary of the D.S.I.R., Dr E.Marsden. While Marsden was naturally reluctant to have his views quoted in an application to the Minister of Finance for the funding of a new course, he had no objection to offering his private opinion. "It is the best thing that can be done for Industry in New Zealand. The one thing that is stopping the development of secondary industries is the lack of adequately trained Chemical Engineers or Managers."[4]

Denham died suddenly at the beginning of 1943, and J.Packer,* who was appointed to the Chemistry chair in 1944, took over Denham's negotiations with the Council, convinced that they were a "matter of considerable national importance, and one of some urgency . . . as a result of the changed conditions resulting from the war." Packer's recommendation for the new course had two components:

1. A four years' course in Engineering Chemistry, leading to the degree of B.Sc.(Eng. Chem.), within the faculty of Science.
2. A one year's post graduate (Post B.Sc. or M.Sc.) course, leading to the Diploma in Industrial Chemistry (Dip. Ind. Chem.)

The advantage of this two-pronged approach was that it shortcircuited "a part at least of the usual, but somewhat time-consuming procedure"[5] as far as the Senate of the University of New Zealand was concerned. The Diploma course and, incidentally, the inclusion of a new subject, Technical Chemistry I, in the B.Sc. course, could be implemented without delay, while the Government and the Senate worked on the proposals for the degree course. In the event, Government approval of the additional grant of £1,100, required to establish Chemical Engineering, was announced in October 1943. The salary for a lecturer in Applied Chemistry was set at £700, which must have greatly relieved Packer, who fully

*Packer, John, 1899-1971; chemist, teacher, administrator. B Melbourne; educ Haileybury Coll, Melbourne Univ (MSc), London (DIC); FRIC, FNZIC, hon DSc (Cant); apptd lect chem CC, 1923; prof, 1944; ret, 1964; dep cm PB, m Cl CUC; m UNZ Ac Bd, 1943-61; Sen, 1954-61; m UGC Research Com, 1961-4, 1965-71.

realised how unlikely it was that a suitable man would be attracted to the post at the earlier suggested salary of £550. He had written to J.S.H.Jones of the Chemistry Department of the University of Queensland, after Denham's death, to enquire about the Applied Science courses at that university. Jones was very uneasy at the prospect of a salary of a mere £550, as he had grave doubts that anyone suitable would apply under such conditions. He advised that it might be necessary to establish a Chair in Applied Chemistry and also offered a warning about which faculty should "own" the course in Applied Chemistry: "I hope if any . . . trouble arises in N.Z. the course will *not* be put under Engineering control as they have no realisation of what is wanted in the course — the men must be chemists first and Engineers afterwards."[6] So Chemical Engineering was established under the supervision of the professor of chemistry, and space — although very little — was provided for it in the basement of the already overcrowded Chemistry Department. It was not until the move to Ilam that Chemical Engineering became a separate department with allegiance to Engineering rather than Chemistry, and then the reasons were pragmatic, not necessarily a re-thinking of the nature of the course offered.

The one year post-graduate diploma course was established in 1944, and the Senate approved the five year degree course, which commenced in 1945. The problem of finding a suitable man to co-ordinate the new course exercised the Council for some time. As Chemical Engineering was a relative newcomer to the scene, there were few qualified men in Australasia who would be capable of teaching and indeed creating a course in Industrial Chemistry. Once the decision had been reached that the new course would make its début from the Chemistry Department and that the engineering components would be provided mainly by Steele and the Mechanical Engineering Department, it meant that the Council was looking for someone whose main area was chemistry. However, what was essential to the smooth merging of the new course with both Chemistry and Engineering was that the personality of the new teacher should be such as to ensure a good working relationship both within the Chemistry Department and with local industrial firms who had given so generously of their support and encouragement, through gifts of equipment, money for scholarships, as well as instituting a memorial fund to Dr. Denham. S.Siemon,* at that time teaching at the Brisbane

*Siemon, Stanley Robert, 1916- ; b Australia, educ Univ Queensland (M.Appl.Sci.) 1941, Univ Melbourne (ME) 1966; lect CU, 1944, prof 1957-61; first prof Chem Eng, Melbourne, 1965-76.

Technical School, was "the only applicant with the kind of training and experience" that was required, with "very good academic qualifications."[7] But the Council's anxiety about the new applicant's "personality factors," led Packer to write to the former physics professor Dr F.W.G.White who had been seconded to the Australian C.S.I.R.O. for essential war work. White was asked if he would interview Siemon to ensure that he was not the "type of man who gets up against everyone or who can't resist talking down to the practical man, thereby alienating him."[8]

The report on Siemon was satisfactory and his appointment was secure. His subsequent performance completely justified the appointment. It says much for the personal interaction and tolerance of both Packer and Siemon that they were able to survive amicably over the years, when Chemical Engineering added so greatly to the problems of overcrowding in the Chemistry Department. Packer was able to state in his annual confidential report for 1946, a year after Siemon's arrival in New Zealand, that he had given him "a very free hand in the development of his work . . . He has shown himself a good lecturer and teacher and is very keen on both the theoretical and practical aspects of his work."[9] Siemon lost no time in making valuable contacts with industrialists and chemists in the main centres of New Zealand.

Certainly Siemon was an acquisition. He encouraged research in his mini-department, pursuing his own research into comminution of solids; in 1948 he was offered a chance of a two year course at Birmingham University, but was unable to accept the offer. The Board allowed him leave of absence on condition that a suitable replacement could be found but none was available. However, after Dr.T.Hagyard was appointed lecturer of Chemical Engineering at the end of 1949, it became possible for Siemon to take up 15 months' refresher leave. At the end of 1951 he was awarded a Nuffield Dominion Travelling Scholarship, as well as a three months' Carnegie grant for maintenance which enabled him to travel back from his year in England by way of the States.

The Council's eventual decision to establish a Department and therefore a chair of Chemical Engineering meant that Chemical Engineering moved out to Ilam at the same time as the rest of the School. Such a decision was partly economic, as the Council expected thereby that the Government would assume financial responsibility if the Department of Chemical Engineering was subsumed under the umbrella of the School of Engineering.

Chemical Engineering, from its inception, received immense support from industry, not just in the form of encouragement and appreciation, but in donations of plant and money for scholarships. Denham's part in founding the new department was not overlooked when the School of Engineering was built at Ilam. There is a Denham Chemical Engineering Laboratory in his memory.

CHAPTER TEN

Electronics

Electronics was established at Canterbury, again seemingly on the strong recommendation of Sir John Madsen, after his visit in 1943. But it would be closer to the truth to state that Sir John was invited to look over the School because he was known to have a specialist interest in this branch of Electrical Engineering, often referred to as Communications Engineering to distinguish it from the more traditional Power Engineering. As Head of the Department of Electrical Engineering at Sydney University, Sir John had professional associations with the Canterbury physics professor, Dr F.W.G.White, as well as with Tom Pollard. It would appear that Pollard's concern with the quality of engineering teaching offered at Canterbury involved him in a great deal of backstage activity at this time.

Early in 1939, as war became inevitable, the Air Ministry established an Electrical and Wireless School at the Royal New Zealand Air Force Station at Wigram and Pollard's services were required on a part-time basis. He was obviously most enthusiastic about such an undertaking and assured the Council that "this work will in no way effect (sic) my Academic duties," as he would be required for not more than six hours a week. "Further, from my close association with this, the Wireless Section of the R.N.Z.A.F., I will gain a great deal of useful knowledge in the Radio branch of Electrical Engineering."[1] In fact this was the beginning of Pollard's highly successful career in Electronics which was to take him on a tangential course away from the regular teaching of the Electrical Engineering Department.

The first sour note was sounded with Powell's staff report of October 1939. In his opinion, Pollard's Air Force work had "not interfered greatly with his College work," although the whole "position of Mr Pollard will require very serious consideration." Powell was facing the unpalatable prospect of losing his most dedicated and inspired teacher because of Government war needs, a loss that "will prove very inconvenient to the School of Engineering."[2] New moves were in the air. Dr White had been invited by the Government through the Director General of the Post Office, "to arrange a course of training for radio-physicists in the Physical Laboratory of Canterbury College"[3] and was requesting Pollard's services. The two men had already been engaged on radar research at this time. Powell had reluctantly to accept that he would have to find a replacement. He was remarkably fortunate in that R.D.Neale, who was to prove another dedicated teacher, had but recently returned from his year's stint as an exhibition holder with the British firm, General Electric Company. Initially he was part-time but, in 1942, was made full-time. Increasingly, the Radio Development Branch of the D.S.I.R. occupied all of Pollard's time and energies so that he was granted leave of absence from his College duties for the duration of the war. At first his title was Engineer in Charge but ultimately he was styled Director of the various related branches of the Defence Development Section of the D.S.I.R.

Pollard was obviously in his element; he throve on hard work, long hours and responsibility and brought to his secret and important war work — the development of radar equipment — the same energy and enthusiasm which he had previously given to the School of Engineering. However, it was an anomalous situation; the Radio Development Branch had its headquarters and workshop at Canterbury College; it took over much-needed and valuable space in an already overcrowded College; it used working hours of various technicians and eventually overwhelmed the whole of the students' gymnasium. The students, especially the Football Association, regarded this as Government encroachment and resisted the whole manoeuvre as best they could. At first it was only a partial take-over, although even this necessitated the removal of the scrum machine to an alternative location at D.S.I.R. expense. The students' fear that this was the thin end of the wedge was fully justified; shortly afterwards it became necessary for Pollard's Radio Development Branch to requisition the whole of the gymnasium for the duration of the war. It was logical: the space was there; by no means all the students used the gymnasium; the scrum machine had been accommodated elsewhere; the Fencing Club was given the use of the College Hall. But the students were most irate and resented his high-handed manner of appropriating their space. As Pollard saw their futile protests as obstruction of a vital war effort, the whole situation was potentially explosive. (Even after the war, the gymnasium and swimming pool were not sacrosanct. During the inter-regnum before the School could move to Ilam, Powell, without compunction, assigned the gymnasium and the pool to be built over to provide additional engineering accommodation.) A police watch had to be set on the gymnasium, not because of student vandalism, but because the radar equipment stored there was top secret and valuable. At one stage, however, the Radio Development Section erected a delicate radar antenna, not unlike a gigantic television aerial, on the Hereford Street frontage. The next morning it had been unceremoniously converted into a clothes line, and was festooned with knickers filched on a panty raid of one of the women's hostels. By all reports, Pollard was not amused.

By 1942 Pollard was officially employed by the D.S.I.R., but he was far from inactive regarding the School of Engineering and the need to establish an Electronics Department. Earlier attempts to induce the Engineering Faculty to accept a new course had reached a deadlock because of the difficulty of agreeing on the amount of Physics which should be included in the course. Pollard's enthusiasm was boosted by a visit which he made to Australia at the end of 1942; he submitted a detailed report of the information he had acquired on his visit to Sydney. The Council expressed interest and a meeting of the Special Engineering Sub-Committee resolved that Pollard and Powell should confer on the subject. (This special committee was one of those set up to discuss engineering matters following the subversive attack in *Canta* at the end of the previous year. Amongst other items, the members examined the viability of establishing both Chemical Engineering and Electronics.) Pollard had written:

> The chief factor of interest in the Electrical Engineering course at Sydney University is in the special arrangements which have been made there by Sir John Madsen, Professor of Electrical Engineering, in order to broaden the education of the senior Electrical Engineering students, and in order to include in that course a more general treatment of Electrical Engineering with some special reference to modern advancements in Communications.[4]

When consulted, Powell was characteristically cautious about the possible new developments. He queried whether they would be weighted more towards science or towards engineering. However, "the proposal to include more Physics is somewhat attractive. It would imply greater co-operation between the School of Engineering and the Physics department than has been the case in the past, and I should remark that any future cooperation must not be one-sided." (The irony of this remark in the light of Scott's battle to block the establishment of a separate Physics chair was undoubtedly lost on the current members of the Council who would have been ignorant of the history of the Engineering School's relationship with the Physics Department.) However, his real objection was that he was "in general opposed to any specialist course; it should not be the function of a University to train for a particular department."[5] Such emphasis would be at the expense of fundamental engineering training.

Powell's opinion was given in good faith with no suspicion that the Council had come to doubt his credibility. But only a few weeks after he submitted his thoughts on the proposed Electronics course, the confidential letters from recent graduates arrived with their recommendations that both he and Steele should be set aside if the School were to regain its former momentum and reputation. While a beleaguered Council would not consider dismissing out of hand two loyal and long-serving staff members, it was likely to view Powell's cautious response to Pollard's enthusiasm with a jaundiced eye. The School had been receiving a bad press; what was needed was innovation. Electronics, as well as Chemical Engineering, would provide the right kind of publicity. And it was much easier to approve exciting new developments than to conceive how to modify existing courses which had grown stagnant with the weight of tradition. Pollard's enthusiasm and creativity were a god-send; if any Council member recognised the shade of Scott, nothing was said. It was necessary to rescue the School of Engineering from the doldrums, and empire-building was, by this token, acceptable.

The scene was already set, therefore, for Sir John Madsen's visit to the School and for his strong recommendation that the Council should establish a course in Communication Engineering, which was "one phase of the College activities directly associated with the war effort, upon which immediate action can be taken with advantage." An appeal to patriotism was unnecessary; the Council was already well primed. As a temporary measure, all that would be required was "the immediate expenditure of approximately £300;" the "Council should make such a sum available immediately, to be spent directly under the supervision of Mr Pollard, acting in his capacity as Senior Lecturer in the department of Electrical Engineering, the necessary delegation of authority from Professor Powell being made to enable him to proceed directly with the full control and development of this laboratory."[6] Such a recommendation to set up a virtually independent establishment within the Electrical Engineering Department posed a very delicate problem in itself. The problem was compounded still further by the fact that Pollard was currently employed by the D.S.I.R., on extended leave of absence from the College, and therefore no longer accountable to Powell.

In the meantime, the Council agreed to introduce an Electronics course and followed the procedure recommended by Sir John, making a grant of £300 and asking Pollard "to initiate an Electronic Department as part of the School of Engineering."[7] (It is obvious from Pollard's account of this period that he believed that ultimately there would be a separate Department of Electronics under a lecturer in charge. With hindsight we can see that the council behaved equivocally, encouraging Pollard to take full responsibility for the new development, but at the same time withholding from him the necessary status which would have enabled him to have acted freely without the need to constantly refer his decision to Powell as Head of the Department. It seems likely that if Pollard had not been divided between his work for the D.S.I.R. and his commitment to Electronics, the Council might have established a separate department right from the beginning. Much bitterness and rancour would have been avoided. As it was, the initial suggestion from the Council was obviously couched in most ambiguous terms.) The Minister of Education was approached for the necessary authority, which was duly given. Pollard was appointed Senior Lecturer in Electronics in 1944, after he had been asked to resign his senior lectureship in Electrical Engineering. Despite the already heavy load which he was carrying with his war work for the Radio Development Branch, he found the time and energy to publicise the new development. Like Scott before him, he was "a persuasive beggar."[8] He wrote letters to all the leading firms which he thought might be interested, he publicised the new course on all his travels: "The results justified my efforts and by 1944 industry had contributed £2000 in cash grants for the purchase of electronic apparatus to be used for educational work and industrial research. Various representatives of overseas firms allowed me to select electronic apparatus which their principals gave to Canterbury College; the value of this equipment is over £2000. Two scholarships each tenable at £150 per annum were given."[9]

The result of all this activity, however, was that the gap was widening between Electronics and the mainstream of Electrical Engineering. For, Pollard as senior lecturer in Electrical Engineering under Powell's supervision and authority had been one thing; Pollard still nominally in the Electrical Engineering Department but basically sole lecturer in Electronics aware of how much the School was indebted to his energy and efforts for the donations of money and plant it had received, was another. In the past, there had obviously existed the most cordial relations between him and Powell. He had originally been appointed demonstrator in Electrical Engineering in 1922 on Powell's strong recommendation. Although Pollard had no degree, Powell claimed he was a more impressive student than many of his graduate contemporaries. Certainly he received, each year, high commendation from Powell in his annual staff reports. Former students also remember him as a stimulating teacher, a tribute not paid to many of the Engineering School staff in those days. Now, however, there was evidence of a slight rift between the two men, caused mainly, it would seem, by Pollard usurping authority. Letters from the Registrar made it clear that Pollard was assuming that he had the right to make decisions, especially organisational and financial ones, without first getting the go-ahead from Powell. He was writing direct to the Registrar, giving instructions, without informing Powell of his intentions nor gaining his approval. He had to be reprimanded for this more than once. And there was nothing conciliatory about Pollard's tone; he advised, informed, recommended and despite occasional demurs from the Registrar, his requirements were met, on the whole, even if at some inconvenience to others. In a letter to Powell, as Dean, about the storage needs of Electronics, Pollard stated peremptorily:

> I have had a number of talks with you re taking over the cement room, but, so far, little progress appears to have been made in shifting the cement mixing. Apart from the fact that this room will be necessary next year and that a lot of work will have to be done on it, I am very worried about the cement dust and dirt getting in to some of the precision apparatus which is now arriving.

A handwritten note from the Registrar's office at the bottom of this letter commented: "Powell asked to arrange a conference (Dean, Steele, and Pollard) to discuss question of cement and arrive at a mutual agreement."[10] The delicate precision equipment required for the new course obviously had to be stored most carefully. Thus an already drastically overcrowded Engineering School had somehow to accommodate a considerable amount of new equipment. Pollard's attempts to requisition the Mechanical Engineering Department's cement room were far from palatable to Steele, who objected to his own needs being thus lightly set aside. Steele grumbled in vain.

Once the war was over, it was assumed that Pollard would be released from his D.S.I.R. work so that he could devote all his time to the new Electronics section. But this was not to be. The R.N.Z.A.F. had requested four radar Height Finders for meteorological service. As "the electronic and mechanical portions of this structure are original, having been designed and developed"[11] by the Radio Development Branch, his directorship was still needed. Although he was full-time with the D.S.I.R., he continued to give lectures in Electronic Engineering, to provide the necessary instruction and supervision to the two electronic scholarship men, and supervise generally the Electronic Department. His energy appeared to be unbounded. The situation was most unsatisfactory and Powell, as Dean, not surprisingly queried its advisability: ". . . it would be very desirable to obtain from either the D.S.I.R. or Mr. Pollard or both a definite statement as to the state of affairs that will occur next year."[12] As Powell was due to retire, it was doubly necessary that there should be some clear statement of intent before a new professor was appointed.

Actually, the problem of Pollard's divided allegiance — between the School of Engineering and the D.S.I.R. — was not to be lightly or easily resolved. By January 1947 he was submitting to the Council a memorandum from the Secretary of the D.S.I.R., "giving proposals for continuing the D.S.I.R. set-up at Canterbury College." The Canterbury College Industrial Development Department was about to be born. He regarded such "an arrangement . . . [as] of benefit both to Canterbury College and to this department," not just because of the interaction with industry which such an organisation would foster, but because of "the advantages of a completely equipped and manned workshop" for both the Engineering and Science departments. His own role in this new "set-up", as he saw it, was that he would nominally carry on in the position of director, "which would allow [him] the general control of the organisation," with a reasonably senior person appointed to take over the bulk of the general administrative duties:

> The advantages of such an arrangement are obvious, as it means that a College man is controlling the organisation, but, at the same time, he is acceptable to the D.S.I.R. From my own point of view I favour this idea, as it allows me to keep in contact with industrial work, which is advantageous to the Electrical Department.

Canterbury College's access to the workshop was the most important bargaining factor offered by Pollard. After all, "the alternative to this arrangement is for the D.S.I.R. to shift its staff, tools and stores elsewhere, and for the College to set up its own central workshop and employ its own staff. This will involve the obtaining and purchasing of machine tools and the even more difficult question of stores. Assuming the same average salary it would not be possible to operate at [the current rate of] 4/- an hour."[13] By this time, the Council had become accustomed to the presence of D.S.I.R. personnel on campus; to continue the association was obviously going to require less effort and less expenditure than if there were to be a total severance. Pollard's arguments were most persuasive and Powell, who was on the point of retiring, by this stage was out of account. Actually the Council needed little persuading. Anything which would help Canterbury to regain its clear lead over Auckland was welcomed at this stage. (It was at this time that the title, "The National School of Engineering" began to figure prominently on letterheads and official statements. It appears that, as Auckland's School of Engineering received Government recognition so that all branches of the B.E. degree could be taken there, instead of Auckland students having to come down to Canterbury for their final year, then Canterbury became more defensive about its own diminishing prestige. The Government acknowledgement in 1921 that the Canterbury School was "a" national School of Engineering undoubtedly referred to its acceptance of financial responsibility towards an institution which provided an engineering education not just for local students but for students throughout New Zealand. "A" National School was readily transmogrified into "The" National School, but while there was only one School, this was of no account. Once Auckland had levelled out in the race against Canterbury and then threatened to outstrip it, it became far more important for Canterbury to parade its original title, and far more offensive for Auckland that it continued to do so.)

At this stage, Pollard could be assured of the Council's support and approval for his schemes. As well, with Powell's retirement, the chair of Electrical Engineering was vacant and because of a debate about what salary was appropriate, there was a considerable delay before the position was advertised. During this period, therefore, Pollard, in addition to his already onerous position with the C.C.I.D.D., accepted the role of Acting-Professor in the Electrical Engineering Department. The extent to which Pollard's star was in the ascendancy can be seen by the fact that in September 1947, the Engineering Sub-committee recommended that "Electronics be constituted a separate Department of the National School of Engineering."[14] The Council unhesitatingly accepted the recommendation.

N.M.MacElwee of Birmingham University was the successful applicant for the chair, despite "his somewhat retiring nature."[15] MacElwee lost no time in trying to make himself familiar with the running of the Electrical Engineering Department, and even before leaving England he wrote to the Canterbury Registrar with a list of information he required about the department he would shortly be heading. He laid special emphasis on the following items:

> C. A detailed, lecture by lecture, syllabus for each course of lectures giving three or four lines to each lecture and covering subject matter and method of presentation whether descriptive, algebraic, by vector diagram, vector algebra, etc., . . .
> F. A list of the main items of laboratory plant and equipment with a brief note on each item:- rating and purpose for which it is generally used.[16]

MacElwee's letter was forwarded to Pollard, who replied in a manner which can only be described as brusque. In response to MacElwee's request for information about lectures, he wrote:

> I regret it is quite impossible for me to give you this information — neither Mr Neale or myself have made a practice of giving a set schedule of lectures. Frankly, the syllabus as a whole is in need of revision . . . As regards the lectures themselves Professor Powell lent me certain of his lecture notes but they were of very little use as I find it impossible to lecture from set notes.

Pollard's reply to MacElwee's question about equipment was equally grudging: "I am afraid it would be of little value to supply you with a detailed list of the machines in the Electrical Laboratory. The lab. is fairly well equipped with the normal types of machines . . . "[17] Meanwhile the news had reached MacElwee that Electronics had been designated a separate department. As he had believed that he was accepting a position as Professor of Electrical Engineering, including electronics, this information was highly unsatisfactory. He wrote to W.Wilson, a Canterbury College engineering graduate who worked in Great Britain for The General Electric Company, and who had evidently been instrumental in recommending that MacElwee apply for the chair. Wilson recalled the interview MacElwee had attended with the London Committee: "The Chairman definitely enquired from you your knowledge of electronics" and one of the reasons why you were recommended was that "although you were primarily a Power Engineer, you had had valuable experience and were well versed in electronics, and especially its application to heavy power engineering." Wilson also commented: "if anything goes wrong, you may feel safer with the knowledge that your old position is still open." Birmingham University had regarded MacElwee's departure as a "heavy blow"[18] and had not, at that stage, found a suitable replacement. Strengthened by such information, MacElwee approached the Canterbury Council concerning the terms of his appointment. MacElwee's reaction to the separation of Electronics from Electrical Engineering obviously put the Council on the spot. The last thing it wanted was to have its newly appointed professor deciding to relinquish his position. Accordingly, it rescinded the previous motion. At the time, Pollard was in Fiji and on his return requested an interview with the Rector and the Chairman of the Council. He was most irate. He referred to the discussions which had previously taken place and the unanimous concensus at the various levels that "it would be in the best interests of Canterbury College and the School of Engineering if a separate Department of Electronics were set up. Now, without any further reference to me or explanation I am told the resolution has been rescinded."[19]

This was January, 1948. Pollard had the interview which he had requested but the Council was not willing to re-establish Electronics as a separate department. The rift which had opened between Powell and Pollard had never reached unworkable proportions; that between MacElwee and Pollard rapidly became an impossible barrier. By March of the same year, MacElwee was requesting some clarification of Pollard's position: "This is particularly desirable in view of the fact that his nominal title as senior lecturer in my department, combined with his very natural desire to place in the fore-front the interests of the [C.C.I.D.D.] shows a tendency, at the moment, to disrupt the very excellent team spirit for academic advancement, which I find in the other members of my staff." Obviously Pollard was reluctant to relinquish the coveted title, lecturer in charge, and MacElwee could not stomach this. "Mr. Pollard considers himself, within this title, to have independent powers of direction and control. Powers which I should have been only too pleased to deputise to him, in due course, within the general framework of departmental policy, had he shown any co-operation, or even interest, in such matters." The relationship between the two men did not improve. In September 1948, MacElwee submitted a confidential report to the College Committee, which made it clear that he found the situation intolerable. The other members of staff, Neale, W.Bowen and by this time B.Withers had "laid the foundations of a first-class team spirit, which looks to me for leadership." Contrary to this: "although fully aware of his status within the Electrical Engineering Department, [Pollard] has remained . . . the non-cooperating head of a separate organisation in which he includes Electronics, retaining until recently the order book and records of that section, acting without reference to me in important matters both internal and external, and opposing every effort made to secure co-ordinated and harmonious working in the Department."[20] While the Council might well have wished to recommend patience in the hopes that the tension might in time be defused, MacElwee was not prepared to wait indefinitely:

> In relation to these issues it is never-the-less essential that I should state, that conditions have already been produced, which throw doubt upon my personal integrity and militate against the establishment of my professional reputation in New Zealand. The continuation of these conditions will almost inevitably lead to my resignation.[21]

When the dispute was referred to Tocker, the Rector, he reported that "Mr.Pollard says he has always been willing to co-operate, has been on the friendliest terms with Professor MacElwee, has had no complaint from him, and was very surprised to learn that Professor MacElwee had complained to the Council." Tocker also remarked that MacElwee had "advanced some rather vague and formless complaints [which] it would be difficult to substantiate."[22] A special committee was convened to discuss the situation; both men were interviewed separately. It recommended, with Tocker dissenting: "That, (as a reconciliation between Professor MacElwee and his Senior Lecturer appears impossible,) Mr.T.R.Pollard be requested to relinquish his lectureship in the Electrical Engineering Department (to which he now devotes only part time) and that authorisation be given to appoint a replacement." The greatest efforts were made to safeguard Pollard, not just in the recommendation that he be offered the position of Deputy Director of the C.C.I.D.D., but in the even more specific recommendation covering the possibility that, at some future stage, he might desire to return to the School of Engineering, when "every consideration should be given and every effort made"[23] to re-appoint him as Senior Lecturer in Electrical Engineering. Such expressions of partisanship towards Pollard made it quite clear that the Council were asking him to resign under protest, and only because the present situation had become untenable.

Pollard did resign, but there was still the vexed question of how to re-organise accommodation both for the C.C.I.D.D. personnel and its equipment. A temporary arrangement allowed the C.C.I.D.D. to have space in the Junior Electrical Laboratory until the beginning of the following academic year. As a result of a delay in the arrival of insulators for the surge generator, it was possible to extend this privilege until June 1949 but not beyond. A cautious letter from Calvert, as Dean, had various additional notations asking the Registrar to "please confer with Director C.C.I.D.D. and see if this request can be implemented with as little upset as possible."[24] Even so, MacElwee had to complain more forcibly before the C.C.I.D.D. vacated the laboratory: "Compromise on the original agreement produced a concession, and it is now suggested that the concession be converted into a right." He requested that "further delay of my departmental work will not arise at this critical time."[25] As a result of this communication with the Rector, Pollard was instructed to remove all gear immediately. MacElwee was fully justified in demanding space in his laboratory, but there was always such a self-righteous tone to all his statements that it is not surprising that Pollard should regard them as "a vague collection of trivialities".[26] Pollard was not willing to be suitably subordinate. Either Electronics had to be a separate department under Pollard's leadership, or Pollard had to go. As he was already giving most of his time to the C.C.I.D.D., it was logical that he should be the one asked to resign. The whole episode was unfortunate, but quite obviously the personality clash between the two men was unmanageable.

Pollards's insistence on the title, lecturer in charge, is strongly reminiscent of the similar battle Scott had to establish himself in the early days of the School's history. Actually, apart from a different attitude towards their lectures — Scott was always immensely careful that the copyright on his lectures should be safeguarded while Pollard exhibited a total insouciance with his refusal to lecture "from the book" — the similarity between the two men is remarkable: amazing energy and dedication to the cause of engineering education; a forthright, potentially dictatorial manner brooking no questioning of motives; excellent P.R. — both men greatly added to the equipment maintained by the School by winning the enthusiasm of industry towards their endeavours; even the tendency to quote their own earlier letters in support of their arguments. Each of them, in his separate way, had to struggle against the implicit academic condescension which degree-holders showed towards their academically deprived colleagues. Some men, eg. Calvert, were able to withstand such academic pressure; others were especially sensitive to it and reacted forcibly against any attempt to devalue their achievements or their status. In Scott's case, he was having to establish himself within a hierarchical system which, traditionally, had no place for engineering; Pollard was up against an academic outsider, who, with little knowledge or understanding of the situation within New Zealand, was very sensitive that his own rights and privileges should be observed.

Pollard himself must have identified strongly with Scott; he had studied under him and Scott, after all, had become a legend in his own lifetime, at least in engineering circles. Although at the time of Scott's death it was stated that there would never come a day when his achievement would have faded from the memories of all those connected with the School of Engineering, at the time of the move out to Ilam, only Rastrick of the professorial staff had studied at Canterbury, and he had never been taught by Scott. Obviously anxious that Scott was not going to be adequately commemorated in the new building, Pollard wrote to the Council about the ideal placing of Scott's portrait in the new building. He suggested there should be a bronze plaque as well, describing Scott as the Founder of the School of Engineering. The Council was not comfortable with that designation, although it agreed that Scott's portrait should be prominently displayed. The plaque which was provided reads instead: "Head of the School of Engineering."

Pollard was also concerned that Scott's grave had been allowed to become overgrown for want of attention because there were no close surviving relatives to tend it. He asked that the Council accept the responsibility. The matter was referred to the Engineering Faculty and the pragmatic response returned from Siemon, as Dean, that while it agreed that it should be looked after, it would "save the cost of maintenance"[27] if it were to be concreted over. His recommendation was accepted.

CHAPTER ELEVEN

Breaking-Point

While the row over territory was raging between Pollard and MacElwee, the Engineering School was beset with other, more serious difficulties. The problems of insufficient space, obsolete machinery, inadequate staff:student ratios and the complete absence of research had been endured with some measure of equanimity during the war years. Once the war was over, the School was confronted dramatically with the deterioration in its standard of teaching; it was imperative that the situation should not be allowed to continue. The invasion of returned servicemen, who had to be accommodated somehow, exacerbated the problem, especially as any restrictions placed on the numbers of students enrolling could not apply to these men. If the difficulties facing the School had been serious before, they had now reached intolerable proportions. There was no room for expansion on the town site; the College proper was as limited as the School of Engineering in this regard. The complicated manoeuvres which led to the decision to move all the College out to Ilam have been told in detail in *A History of the University of Canterbury*, where, understandably, the spotlight was trained on the College as a whole, not especially on the School of Engineering.

With the prospect of a greatly increased student role as the war ended and men returned to New Zealand, the Council requested Powell to comment on the future accommodation requirements of the School. His reply was unequivocal: "the School of Engineering buildings have reached the limit of their age, they are the result of a series of patchwork additions, the latest development has gone to aggravate this evil." It was impossible for the Council not to concur with such a description. Powell continued authoritatively: "the most satisfactory scheme is one that removes the School of Engineering entirely from the rest of the College buildings, to an area large enough to allow extension as needed."[1]

At this stage, the Council was optimistic. Steele was informed: "it is hoped that the School of Engineering will be rebuilt or at least extensively remodelled within the near future."[2] But this was easier said than done. While the prevailing mood was one which favoured the retention of the town site, it was difficult to imagine how the School's plight could be relieved. The possibility that the School could take over the Normal School was mooted. As Normal was within walking distance of the College, this offered a compromise solution: a new building without relinquishing the advantages of a central position. Doubts were expressed that the functioning of the School might prove too noisy; reassurances were offered. But the main drawback to the Normal site was that the School would be required to take over the existing building; understandably, reluctance was expressed that this would merely introduce another era of makeshift and compromise.

Meanwhile a warning note was sounded, which temporarily limited the enthusiasm of those who were the most vociferous advocates of a complete rebuilding of the School. Canterbury had watched with increasing alarm the rising prestige of the School of Engineering at Auckland. Now it was pointed out that "if Canterbury asks for a new School then Auckland University College will claim that their School should be extended in preference to building an enlarged School in Christchurch . . . Finance for extension of the present School would probably be more readily available than for building a completely new School."[3] Obviously, Canterbury had need to move cautiously. The years of its undisputed precedence over Auckland were well in the past.

But dissatisfaction and frustration were building up. While the Council was exercising itself on the question of whether it was desirable to re-site the whole College, there were strong complaints at the "absence of professional engineers from membership of the Canterbury University Council."[4] When the motion to vacate the town site was decisively lost on 26 May 1947, members of the engineering community understandably felt that their needs for expansion had been totally ignored.

Consequently a sub-committee of the Policy and Development Committee recommended that a special sub-committee "with outside experts co-opted [should] be set up to investigate"[5] the advantages of a move to Burnside — the possible site for a new School at that time was land which later was to be used for Burnside High School. This engineering advisory panel consisted of: J.R.Templin, (a leading consulting engineer), J.C.Forsythe (chief electrical engineer for the M.E.D.),

E.Somers, (city engineer), E.F.Evans (district engineer of the P.W.D.), and H. B.Anderson (managing director of Andersons Ltd.). Engineering support for a move to Burnside was universal. Powell, in his capacity as Dean, the members of the N.Z.I.E. and of its dynamic break-away group, the P.E.A., all strongly favoured the move. But a year later, nothing much had happened other than an endorsement of the need for "the subsequent removal of the School as a whole, until such time as it may be re-absorbed adequately in the College proper."[6]

As sub-committees proliferated, frustration at the cramped conditions within the School was mounting. When the engineering advisory panel brought down its report in July 1949, it was tabled in Parliament. It offered the familiar criticism of the School's overcrowded buildings: "Few, if any, of them measure up to modern requirements, and most of them are crowded, poorly lighted, and inconvenient." It did not favour the adoption of any temporary expedient until such time as it was possible for the College to appropriate more land on the town site. Rather, "something must be done immediately . . . The weight of evidence is clearly in favour of a permanent building on a new and spacious site. The proximity to main lines of communication should of course be kept in mind."[7]

Eventually, in 1950, Government approval was to be granted for the acquisition of the "expensive paddocks"[8] at Ilam. But the tension was not over. The School had to hold its breath and wait for the decision as to which department should be re-sited first, Chemistry having almost equal claim for consideration. The Engineering School's case for priority was based on the School as a self-contained unit, the worry of being overtaken by Auckland, and the importance of engineering to the well-being of the country. Nothing was said about the place of Engineering within the structure of the University as a whole:

> Ever since its foundation the Canterbury College School of Engineering has been recognised as the premier engineering school in the Dominion. It has maintained this position during recent years under conditions of great difficulty due to lack of necessary accommodation. Unless an early start can be made on new buildings for the School, there is a very real danger of the honour passing to Auckland, where adequate accommodation is already being brought into use. Once lost, it may be difficult, if not impossible, for Canterbury to regain the leading position for its special school.

> The future development of New Zealand is largely dependent on an adequate supply of professional engineers, and it would therefore be in the best interests of the country to build a new School of Engineering as early as possible.[9]

Professor Packer put the case for Chemistry. "The staff of the chemistry department is agreed that the advantages to be gained by having adequate facilities for teaching and research in the new building would far outweigh, both for them and the students, the disadvantages of being separated from the rest of the College during the interim period." Inevitably, the place of Chemical Engineering was discussed. At that time, the concensus was that "chemical engineering should remain closely associated with chemistry . . . The existing close collaboration in research"[10] was one of the reasons offered by Packer for Chemistry's retaining Chemical Engineering.

The Professorial Board deliberated. On 7 December 1949 it moved "that, although [it] finds it very difficult to decide, it feels that the best interests of the College as a whole would be served if the first building on the new site were the School of Engineering." The major reason, as was to be expected, was that "the Engineering School is a unit on its own and its move would cause little inconvenience to the rest of the College," and it "could be more administratively self contained than Chemistry."[11] At this stage in the School's history, its finances were mainly separate from the rest of the College, in that it was still receiving a special grant from the Government and did not have to queue for its share with the rest of the College. As well, in 1947, on the recommendation of the Engineering Sub-Committee, the School of Engineering had appointed a Bursar, G.W.Skelton.

The School had been granted priority but there was still a strong feeling among engineers, whether academic or practising, that the move out to Ilam was an appropriate time for the School to sever all contact with the College and assume its own totally independent control. An anonymous article had already appeared in the *Star-Sun* of 8 July 1947 advocating total separation between the School and the College. "It is quite evident that a body whose sole interest was the School of Engineering would be an improvement."[12] The separatism which had been so marked a feature of the School's development over the years was always seen, by engineers, as entirely advantageous to the School. The possibility that the School might profit from a relationship with a University College or with its traditionally established ideals was totally disregarded. Pragmatism demanded that the School be seen as a technological institute, not as one department of a University with no "special" claims for superior treatment. There was no room for sentiment when the future of engineering education throughout New Zealand was at stake. Accordingly, a motion was put forward at the June 1950 meeting of the Council advocating that the School of Engineering should have an independent governing body "to the end that such Board being free of the embarrassments of other interests, will be able better to promote engineering education."[13] Implicit in this statement is the hope that, once freed from the encumbrance of the rest of the College, it would be possible to augment a more favourable salary scale for engineers. The possibility of establishing a differential salary scale for engineers was becoming increasingly remote, for although "the acute situation in the Engineering School could be dealt with by having a special scale for Engineers . . . the problem is one which affects the whole of the academic staff of the College."[14] In the end the motion in favour of separation was lost. The School had to accept that "for better, for worse," it was wedded to Canterbury College; divorce was impossible.

The School's administration was not unjustified in its belief that the Council had failed to appreciate the reality of the situation. Despite the impressive liturgy of complaints it had received over the years about the difficulties of attracting new staff in all departments — sometimes despite several enquiries, there were no actual applicants — the Council maintained in February 1951 that "the Engineering department does not want existing vacancies filled," that "the Engineering department has been liquidating its deficit by drawing on salaries money."[15] This veiled suggestion of misappropriation of funds excited MacElwee's ire. The accusation was "entirely erroneous"; the continuing staff vacancies were due to the lack of suitable applicants, and "the present staff instability made it unwise to proceed with the full development policy."[16]

It was not long before the question of Chemical Engineering, and whether it should retain its allegiance to Chemistry or opt for official standing in the School of Engineering, once more came to the fore. As the engineering components of the course were taught in the Engineering School, — which produced what Siemon referred to as "an unhappy compromise" — Chemical Engineering students were faced, after the move to Ilam, which at that stage no-one anticipated was so far in the future, with the prospect of several years of commuting between the two sites. Apart from this disadvantage, there were other factors to be considered. Currently, research was impossible, "there [being] at present available neither space, time nor equipment. For teaching there is but one totally inadequate basement laboratory, dirty, poorly lit, cold, damp and with a low ceiling. Equipment is sparse, makeshift and developing at a slow rate." After a conversazione in 1953, a writer in the *Press* was surprised that the Council was not ashamed to admit the general public into this basement laboratory.

The Council, in April 1951, requested Siemon to submit a memorandum of his assessment of Chemical Engineering's position. After expressing his appreciation of, and indebtedness to Professors Packer, Calvert, and Rastrick, he stated firmly his conviction that Chemical Engineering should become a full department within the School with its own chair. In this way, Chemical Engineering would benefit from the increased space and the "more liberal treatment of the engineering grant," both of which were necessary if the subject were to advance. "The publicity and prestige attaching to a chair would enable funds for equipment and research to be raised outside of the usual grants." Dr.H.N.Parton, Associate Professor of Chemistry and Acting Head of the Chemistry Department during Packer's absence, strongly supported Siemon's recommendation. "The association of Chemical Engineering with Chemistry has been a happy and fruitful one, and of great mutual benefit . . . [Mr. Siemon] has probably gained valuable experience, widely available in the Science Faculty, in the art of making bricks without straw. He has fully earned a transfer to a better endowed environment . . ."[17]

The Council resolution to establish "a Department of Chemical Engineering within the framework of the School of Engineering"[18] was communicated to Siemon in June 1951, with an expression of gratitude for his continuing good work, but the question of the chair was deferred. It was not, indeed, until the end of 1956 that Siemon was eventually appointed Professor of Chemical Engineering.

CHAPTER TWELVE

The Long Wait

The delay in establishing a chair of Chemical Engineering was nothing beside the delay before the School could occupy its new territory out at Ilam. For, even after the Government had agreed to the purchase of the land and the Council had accepted that the School should take priority, it was not to be until 1959 that the new building would be partly ready to receive staff, students and the massive equipment of the different laboratories. In the intervening years, the School, as it were, held its breath. Even if there had been sufficient money for new developments, there were other hindrances in the immediate post-war years. As a desperate measure to deal with some of the overcrowding at the College, the Council had agreed to the "temporary" erection of steel huts. The School of Engineering was allocated two steel huts of 970 square feet, and one wooden hut, 936 square feet, which put up the total available floor space to 46,176 square feet. But necessary alterations to the old workshop so that it could house several staff members were delayed interminably because of the acute position pertaining in the building trade. In the end, there had to be special submissions direct to the Minister of Housing before the necessary alterations could be authorised.

The School buildings continued to deteriorate, but the Council felt that "major expenses [were] not justified at present."[1] In fact, by 1950, the School was running on an overdraft of £6,000. In such a case there was no lee-way for the kind of luxury advocated by Alley after a particularly fog-bound surveying camp at Godley Head. "A more secure camp in a less foggy district would be desirable . . . with a permanent hut."[2] His hopeful submission was turned down.

Despite the immense difficulties the School was labouring under, however, development was not totally at a standstill. There was definitely a new feeling abroad. One of the major indications of this change in outlook could be seen in the encouragement given to staff members who wished to pursue individual research projects. In the early days of the School's history, testing had taken the place of research. What was needed in colonial New Zealand was an accurate description of basic materials: timber and stone, the iron ore in Taranaki sand. Such activity absorbed time and energy, and initially was seen as entirely justified, a valuable contribution from the School of Engineering to the local community. However, as the years passed, the absence of research, as analysis not description, was frequently deplored. "Professional competence, that is, if it has any meaning beyond acquaintance with the baldest tricks of a trade, is allied with an indispensable closeness to the disinterested pursuit of knowledge."[3] For all the concern increasingly expressed at the absence of research, it took the war, and the technological advances which it produced, before engineering research was regarded as a necessary function of the School.

Active support and encouragement for research projects came partly from the establishment of a University of New Zealand research grant. Siemon was a recipient of such a grant in 1947. Increasingly, as well, staff were allowed financial assistance to attend conferences in Australia as well as in New Zealand, where research papers were read. Then, in 1949, the Air Department presented the School with a £2,500 grant for research into jet engines. This grant was repeated in succeeding years. A new appointee to the Mechanical Engineering staff, H.Adams, was placed in charge of the project. His work certainly prospered and in 1954 he was promoted to the position of Associate Professor. P.J.Alley was another very active researcher — his field was soil mechanics — although the Council ignored Calvert's recommendation, in 1950, that Alley become Associate Professor. Both Adams and Alley contributed articles regularly to the New Zealand engineering journals. This was the tentative beginning of the full-scale research undertaken by the School today.

Engineering students were not immune to the progressive spirit which was developing in the School at the prospect of its move to Ilam. MacElwee's annual report to the Senate in 1954 highly commended the initiative of members of the Engineering Society and especially the leadership of its chairman, D.R.Teplitzky. "During the year the Society sought and obtained permission to make an appeal to engineering firms to provide funds for post-graduate scholarships in engineering."[4] Three firms had responded favourably: New Zealand Industrial Gases, Todd Motors and Skellerup Industries. This brought the number of scholarships available to engineering post-graduates to twelve, of which six were tenable overseas.

There were other indications that the School was picking up momentum again. Calvert had resigned in 1950, believing that "the work of planning the building of the Civil Engineering Department should be undertaken by a Professor who will be in charge for some years after their completion, not by one who is about to retire."[5] Hopkins, who replaced him, no doubt chafed at the long delay before he could enjoy the new department, and received with cynicism the reported reassurance of R.M. (later Sir Ronald) Algie, the Minister of Education, that "the matter [of the rebuilding of the School] was continuously in his mind."[6] Hopkins lost no time in establishing himself as a determined man, stating clearly what his requirements were and not wasting energy on fulsome or extraneous matters. He wrote to the Registrar about a minor change he was proposing: "If you agree with this interpretation, then my action is a purely administrative one, if you don't I should be glad to have the Finance Committee's approval of it."[7]

Hopkins took over at a time when the staffing problem in the department was most acute. His first memorandum to members of the Engineering Committee was headed "Crisis in Civil Engineering." The stage had been reached where it was not just necessary to restrict student entry but there was the added threat that the teaching of hydraulics might have to be suspended altogether. As all engineering students, at that time, were taking hydraulics, this would have meant that there would be no graduates from the School. The crisis was weathered by a tremendous rallying of support from lecturers in other departments, occasional lectures or series of lectures by practising engineers, and a great burden of reponsibility on the shoulders of a young and untried appointee, P.Giddens.

By March of 1952, Hopkins had submitted a second memorandum, this time entitled "Progress in Civil Engineering." Permission had been given him to extend the existing Civil Engineering Laboratory "so that a concrete laboratory of limited area [could] be established." As well, with the co-operation of Professors Rastrick and MacElwee "together with a not wholly satisfactory rearrangement of the Surveying course" it was possible to create a larger Soils laboratory and a Structures laboratory of limited scope. Faced with possible financial stringency, he was incisive that there should be a minimum of compromise:

> Following a recommendation from the Engineering Committee, I submitted on 16th November 1951 a statement to the Finance Committee showing that all unexpended departmental moneys would be spent in equipping these laboratories and that a grant of £4000 would be necessary in 1952 to continue expansion in these fields. Due to the College's severe financial straits this figure was reduced to £3000. I have taken the view that there should be no reduction of standard consequent upon reduction of this grant. Consequently I have instructed the Civil Engineering staff to implement the academic policy determined for the department and that should the funds granted be inadequate I shall make application for a supplementary grant. If such an application is made and cannot be met, the scope, but not the standard will be restricted.[8]

The additional £1000 was granted, which fully vindicated Hopkins's determination that his department would not be forced to mark time until the move to Ilam. His assertiveness earned a sympathetic response from the Council. Apparently, however, MacElwee, as Dean, was less charmed. In 1955, there was a proposal concerning the Civil Engineering tank room which caused a minor crisis, as the proposed conversion encroached on space previously occupied by the duplicating machine. MacElwee complained to Hopkins that his behaviour was high-handed: ". . . it is my experience in the past that you have from time to time taken the unusual stand that a statement made by you in writing which has not been questioned, is accepted as fact by the recipient."[9] The reply was prompt: "I do not accept either your interpretations or opinions as expressed in your letter, but can see no useful purpose in answering them in detail."[10] In opposition to the rest of the College, the School of Engineering presented a united stand. This did not preclude individual rivalries between departments. In later years, an apprehensive Vice-Chancellor setting off to interview "four implacable"[11] engineering professors, was advised to set them one against the other. The old adage "divide and rule" proved as effective as ever.

The history of the School of Engineering falls readily into three phases: the first, from its inception until Scott's retirement spanning a period of 36 years, as Scott himself said, "of an exceptional character"[12]; the second, starting in 1923 and culminating in the move to Ilam in 1961; and the third covering the years of technological expansion up to the present day. The first period, as we have seen, could be characterised by adjectives such as exciting and adventurous; the second period, by contrast, deserved terms such as heavy and frustrating, even inert. While the ruling triumvirate which replaced Scott could not compete with his energy and dynamic forcefulness, prevailing circumstances also made it increasingly difficult for the three professors to initiate sweeping changes. Of these circumstances, some, like the depression and World War II, were economic and political disasters on an international scale; others, like the cumbersome burden of the system of external examinations, required a dominion-wide consensus before they could be removed; others again related to the accommodation problems of the College, with the "patchwork"[13] additions and gradually disintegrating buildings, inadequate lighting and consequent strain and overcrowding; finally there was the "caretaker" nature of the appointments to the chairs of engineering during the middle period, and the vexed question of staff appointments in general when work conditions were so unfavourable and salaries so insufficient.

But now the old School of Engineering, the "rabbit warren," with its wooden stair treads "hollowed by the wear of constant feet"[14] was about to be vacated. Memories of the exciting years of Scott's rule or of the increasing inertia of the middle years were to become vague and insubstantial. A completely new era was beginning; the School, in its greatly enlarged space at Ilam, could once more deservedly claim the position of a premier institution within New Zealand.

Part III: A Terrible Beauty

CHAPTER ONE

The Long Haul to Ilam

The forecast for 9 May 1958 was for cloudy and cold weather with scattered showers. It was optimistic. Heavy rain swept over a crowd of more than 500 gathered that afternoon at an almost empty building site in Creyke Road to listen to an hour of speeches and to watch the Governor-General (Viscount Cobham) lay the foundation stone of the new School of Engineering. Overcoats, a reporter noted, were soaked, water trickled off hats and down necks and reporters' notes became sodden;[1] but the crowd remained. One need not look far for an explanation. The ceremony was not merely the first step in the transfer of a university from its historic central city site, elaborated with cloister and quadrangle, pinnacle and tower, to a spacious if stark new campus. The move was also crucial to the wellbeing of the University. If the engineers did not move to Ilam at the earliest possible opportunity there would be no space on the city site for the fast-growing student rolls in all other faculties. Moreover, the move itself had long been in dispute, one of those protracted and public Christchurch controversies which demonstrate, perhaps more than anything else, the value of the bumbling, democratic process in contrast to would-be planning efficiency. For the academics assembled the occasion was also sweet: the staff of the School, with the help of the profession, had beaten off an attempt to impose a serious limitation on their endeavour. And the School, though it was yet to be built, was symbolic of a new period of expansion within New Zealand after almost a decade of austerity.

But as they listened to Lord Cobham, with his polished urbanity, claim that the British race possessed the qualities of mind and spirit needed for engineering, few realised that they were also witnessing the start of a fundamental change. Ilam served, as one Vice-Chancellor observed, as the outward symbol of an inner metamorphosis.[2] The college, still held in a sort of colonial tutelage to the federal University, had already scented that change. Later that year Canterbury would assume the title of "University" even though it would have to wait for another three years for the substance of change. "Devolution" had become the fashionable word in academia as the University of New Zealand entered its final years; there was a belief that substantial reform would soon transform higher education;* it was becoming clear that at long last research would be accepted as an integral part of the university endeavour; and the Government, having somehow dealt with the post-war baby boom as it erupted in turn through primary and secondary schools, now reluctantly faced the task (and cost) of providing university places on a scale hitherto unknown.

Staff too quickly came to realise that New Zealand, faced with a backlog of essential works delayed by the war and subsequent austerity, required more than the handful of engineering graduates who emerged from the School in the past. Indeed only one factor seemed to be but dimly comprehended amongst all the change. The war had set the world on a course in which science and technology quickly gained an impetus. The knowledge, expertise and imagination of graduates were required as never before and created a demand on a scale unknown before. The engineering professors of the early post-war period made their enrolment forecasts on population trends with reasonable accuracy, but almost every one proved to be conservative because the influence of science and technology was not fully appreciated.**

If it is men, not walls, who make a university, without walls it can scarcely exist. As "rehab" students, and then their children, swelled enrolments at Canterbury, the walls threatened to crumble. The College block was of only five and a half acres and the graceful stone buildings there (the last was built in 1923) had been designed for students in hundreds, most part-time,and not thousands, largely full-time. After 1945 the roll rose to more than 2500. The College Council's decision of 1946 to develop building plans for the eventual full use of the College block as well as the whole of the block to the north was not to stand for

*It came from the recommendations of the Hughes Parry Committee, whose report after its inquiry in 1959 is regarded as a crucial document in the history of New Zealand higher education.

**Their estimates of enrolments were more soundly based than some other predictions. For example, they described statements that the population of Auckland would equal that of the rest of New Zealand by 1975 as "surely fantastic".[3]

long, even though a proposal to build on a near-rural site of 59 acres, on part of which Burnside High School now stands, was defeated.[4]

An unidentified civil engineer appears to have been the first to see the possibilities of moving the School of Engineering, despite temporary disadvantages, to a site outside the city centre. He was quoted as saying that a long-sighted policy would be to recognise that the ultimate needs of the College could not be met even by two city blocks of land. Space was required and that could be found only outside the city. The School was self-contained and could with great advantage be shifted immediately.[5]

That was a public expression of what many engineers were feeling privately because the alternative was an absence of any sort of expansion in a building in any case unsuited to engineering. And other members of the College saw considerable advantages too; with Engineering removed there would be some relief from the congestion being felt on the city site. So it was that a recommendation in the College Committee report at the end of 1948 was approved by the College Council. It gave both the Engineering and Policy and Development Committees authority "to investigate the acquiring of suitable land for the immediate extension of the laboratories and perhaps for the subsequent removal of the School as a whole until such time as it may be re-absorbed into the College proper".[6]

It is said that the first person to foresee the advantages of the Ilam site was P.J. Alley. He drew it to the attention of C.W. Hamann a consulting engineer and member of the University Council (1948-51). The first official mention of the Ilam site was a letter from Hamann to the Council Chairman, Sir Joseph Ward, on 20 January 1949. He had accompanied the first (and new) full-time Rector, Dr H.R. Hulme and members of the School on an inspection of the broad and open fields at Ilam and wrote: "It might prove suitable for the School."[7]

Might? It certainly would. The Department of Education, which continued to maintain detailed control of College affairs despite the establishment by the Senate of a rudimentary University Grants Committtee, had suggested to Auckland, Victoria and Canterbury that they consider moving to new sites away from the city centre. As a result there was support in Wellington as well as among the engineers in Christchurch for leaping into the unknown. It now remained to convince the Council.

The University historians saw the acquisition of the Ilam site as a makeshift, an improvisation;[8] and indeed a resolution before the Council in February 1949 merely proposed that it consider buying an extensive area of land on Clyde Road and removing the School of Engineering to semi-permanent buildings on the new site.[9] But if that seemed all too tentative, later events rapidly hardened the decision. Mr T.H. (later Sir Terence) McCombs,*then Minister of Education, visited the site followed by the Grants Committee. The Council, in the following month, approved a proposal to acquire 133 acres of land (the Gardner block of 49.5 acres west of Clyde Road, the Neave block of 30.5 acres behind it stretching to Ilam Road and the 53 acres of the Stead block between Ilam and Waimairi Roads.[10] And taking its courage in both hands it also asked the Government for a grant to build a new School, to adapt the old School for use by other departments, provide a new heating system in the College and, for good measure, to meet the deficit on the 1948 accounts as well as the projected deficit for 1949.[11]

McCombs said that if the idea was approved, future permanent buildings for the College would "probably" be erected at Riccarton, but the Prime Minister wished to make an inspection.[12] Peter Fraser duly arrived, looked over "a magnificent site" and then heard the case for "extending" the College there. (He also heard a rebuttal by a former chairman of the Council, J.H.E. Schroder**.) who had been the prime mover in the decision to extend the College in the city).[13] Unexpected support came from the report of the Consultative Committee on the Education, Training and Supply of Professional Engineers tabled in Parliament. The building of a new Engineering School in Christchurch was considered of greatest urgency, it said; steps should be taken to secure a suitable site; and the building should be given a high priority.[14] Two months later McCombs announced that the site had been made available for future extension of the College.*** Ward revealed that

*McCombs, Terence Henderson, Kt (1975) OBE (71), ED, 1905-82.; Educ. CBHS, CC; M.Sc.(hons), ARIC, FNZIC; teacher ChCh Tech.Coll. Lytt.DHS, Seddon Tech., Auck; M.P. Lyttelton 35-51, Finance Under-Sec 45-47, Min.Educ. & D.S.I.R. 47-49; head Cashmere HS; Mbr Un.Cl. 36-47 and 57-73, Chancellor 68-73; N.Z. High Comm, London 73-75.

**Schroder, John Henry Earl, (1895-1980). Jnlist.tchr, Asst Dir. & Dir. NZ B'casting Service 49-61; b.Hokitika; educ. CC (M.A.1920); teacher CBHS, asst ed ChCh Sun, asst ed. Press, mbr BG 1929-46, chmn 1944-46.

***In the following year a new Government authorised the purchase of two properties, the Stead block for £32,000 and the Neave block for £11,144. Part of the Gardner property was bought by proclamation in 1951 and the residue in 1967 at a total cost of £156,000. The establishment of the University of Ilam soon affected property values in the area.

the ultimate aim of the Council was to move all departments to Ilam "and on no account should there be a permanent division".[15]

There seemed to be little doubt that an Engineering School would have priority on the new site. Before the outbreak of war, plans had been drawn for a new Library, but by some sleight of committee hand the library was superseded by a new Chemistry building on the site of the Students' Union and test bores had already been taken. There was a long debate in the Professorial Board. It plumped by 16 votes to six for a new Engineering School[16] and the Council agreed, though not without a casualty. Professor R.S. Allan (Geology) resigned, saying he dissented from the priority and should be replaced on the Council by a professor who could give enthusiastic support to the policy.[17]

Why engineering? The Council said in a statement justifying its decision that the School was a unit and could be transferred completely. The move would enable the College library to expand. In the overall plan for Ilam, the School could be sited more independently than the Chemistry Department. And, unlike students in other disciplines, engineering students, at least those in the professional years, would undertake both lectures and laboratory work at Ilam. "From the viewpoint of administration and finance the Engineering School is the simplest unit to move to the new site. Finance for the School is provided separately from the finance for the rest of the College, about 15 per cent coming from student fees and the rest from a Government grant made directly to the School. Only within the School itself does the Council have power to allocate funds."[18]

It is tempting if unprofitable to speculate on the course the Council would have taken had it known that more than a decade would elapse before the School would be occupied. The first National Government, elected on a policy of retrenchment, scarcely deviated from that policy, despite an unexpected and extraordinary boom caused by the Korean war. Higher education fared poorly because such resources that did go into education went first into primary and then secondary schools. The Government bought two blocks of land, but then, to all intents and purposes turned its back on the College and precipitated a period of unprecedented frustration. It was not simply the irritation occasioned by inadequate buildings, salaries and all other resources; there was an air approaching hopelessness as searching schedules were drawn up, plans drawn and discarded, policies approved and rescinded and arguments won and lost. One of the latter was, in fact, partly responsible for the resignation of Hulme and a division on the Council and between the Council and the Engineering staff that threatened to split the College apart. The irony was that despite the long years made available — by inaction in Wellington — for preparing policies and plans, the new School would be too small by the time it was built.

But, perhaps fortunately, the Council had little notion of the difficulties ahead. Many of its members knew S.G. (later Sir Sidney) Holland well; but if they hoped their friendship might be used to advance the College's objectives, they found the new Prime Minister discouraged too close a familiarity with past associates.*

The Council sent its Registrar, James Logie, to Australia to investigate university buildings; a topographical survey of the new site was made; authority to sink test bores was sought; and a general layout plan was prepared. Indeed, enthusiasm outran prudence. Preliminary sketch plans for the School were authorised without the prior approval of the Education Department, which quickly revealed the rigorous nature of the Government's parsimony. A Christchurch architect charged much less than the scale fee in anticipation of doing the detailed planning, but the Department refused to pay any more than £1000 of his fee of £1390. The Director, Dr C.E. Beeby added: "The Minister has asked me to inform the College Council that in future it will not be possible to consider claims for expenditure incurred without prior authority."[19] This was the first dispute involving architects. It was not to be the last.

The first layout plan showed the School fronting Creyke Road. Rastrick, who was Dean, suggested it should front Ilam Road and back on to a wooded area north-west of Okeover (the site of the present Rutherford building for Chemistry and Physics.[20] That idea was unacceptable, but his proposed "ideal arrangement" for the School was adopted and remained remarkably unchanged throughout the next seven years of planning. That arrangement was a core block containing lecture rooms, drawing office, staff rooms, library and administrative offices with laboratory wings leading off "in such a way that each wing is a separate entity with a separate entrance and so arranged that individual

*Not long after his election Holland chided some Christchurch friends who greeted him as "Sid" when they waited on him to press the case for declaring Harewood an international airport. "Look here you fellows", he said stiffly, "I hope you realise that you are now talking at a pretty high level."

development is possible."[21] The Council liked the idea well enough and it began detailed planning.

In the best of all possible worlds the new School would have been built and equipped to cater for those who would enrol over perhaps the next 20 years. The world however is not perfect and the School which finally emerged was too small on the day it opened, necessitating a series of extensions that in total cost almost as much as the original building ($1,390,200 against $1,718,962). Planning might not have been a strength of either engineers or Council, but even if it had been perfect, seemingly capricious orders from the Education Department, and political pressures, would have prejudiced it. Auckland was no longer "that bastard institution," in Scott's derisory phrase, but an engineering school in its own right, even if it had been rusticated to draughty hangars on a disused airfield at Ardmore. No longer did final-year Auckland students come to Canterbury to be finished, as it were, for proper engineering society. Auckland had won that battle and was claiming its prize.

If planning in principle seldom works well in practice, no building could be authorised without a mass of supporting evidence to justify it; and the Council could not hope to avoid becoming closely involved in detail. In the end the recipe used produced a cake of sorts, but too many ingredients were introduced, too many cooks had a hand and the baking time was unconscionably long. The engineering professors provided the major ingredient: past and present enrolments.

A dash of projected enrolments was added after a dutiful look at the post-war birthrate and another rising agent was the report of the Consultative Committee on the Education and Training of Engineers.*

Some provision for incorporating chemical engineering, then part of the Chemistry Department, was thrown in together with an estimate of how the plans of the Institution of Engineers would increase enrolments.**

*Appointed by the Minister of Education (McCombs) in 1948, the Committee recommended that the proportion of engineers in New Zealand be increased to 850 per million of population over the following decade and to 950 in the subsequent 15 years. This compared with 1800 per million in Canada (1941) and 2250 in the United States (1946). The Committee's purpose was to ensure adequate opportunities were given to all persons likely to succeed as engineers. Its estimates of supply and demand began with a shortage of 348 engineers, but it was expected that if its recommendations were adopted the shortage could be reduced to four some 23 years later.[22] Fine tuning indeed!

**The Institution wanted to end what it saw as weaknesses in engineering education by making entry to the profession open only through a university engineering school and not through the associate membership examination of the professional institutions in Britain.

To that general mix was added a generous portion of baking powder — a "multiplying factor" of 1.6, a figure devised by the Education Department for new buildings as a gesture towards meeting the growth of population.

The engineering professors laboured and the outcome was a 37-page report, which based its recommendations on a figure — accepted in the past — of 400 students in their professional years at both engineering schools. They magnanimously and unanimously proposed a 5:3 ratio: 250 for Canterbury and 150 for the other place. Multiplication of that 250 by Beeby's 1.6 gave 400, to which should be added, say, 30 places for graduate students.

The report accepted the Rastrick layout for the School and proposed a minimum of eight lecture rooms, compared with five in the existing School. The School's staff establishment in 1951 was 31. The report conceded that "three or four" would be needed for chemical engineering, sought a room or two for visitors and concluded that 36 staff rooms, the majority of them being of 180 square feet, would be required, together with four professorial suites each of 450 square feet. With laboratories, drawing office, a faculty room, library and administrative offices*the total area proposed was nearly 100,000 square feet.[23]

They were modest enough requirements and Hulme rather chided the authors of the report. "I do not think 31 staff rooms and four suites will be fully adequate," he commented, adding that while he did not believe the staff:student ratio was a very important statistic "it is reasonable to budget for some increase in staff to cover such a large increase in students" — an increase from the current 280 to 430. In addition it was planned to introduce more tutorials, which would require more staff, and there was a possibility that engineering might become a five-year course, as in America. "For these reasons I think we should budget within the confines of the building plan for a staff of 45 in all," the Rector said. But he thought they could be fitted into six rather than eight lecture rooms and he was not convinced that a department's staff should be all together. "Nothing will be lost by mixing various departments up a little and this would have the added advantage of one block of staff rooms not being regarded as the property of a particular

*The professors justified retaining the administrative office by saying the School worked as a unit and is a special School of the New Zealand University. "It should include a general office of 750 square feet with accommodation for three typistes (sic), a Bursar's office and a Dean's office."

department"[24] — a sentiment that did not endear the Rector to any engineer. Worse was to follow.

The College Council discussed the report, slightly amended, and decided to employ the Government Architect to turn proposals into plans after some indiscreet discussion about the abilities of private architects to handle such an assignment. There was a sharp letter from the New Zealand Institute of Architects asking on what authority the statements quoted in press reports were made.[25] The Canterbury branch made a similar demand; and so too did an old friend, R.S.D. Harman, a respected architect who had, with the town planner Nancy Northcroft, given valuable assistance in laying out the site. It would be helpful, said Harman, more in sorrow than in anger, if ill-informed and inaccurate speculation about the architectural profession were not broadcast in the public press.[26] Poor Logie wrote the most diplomatic of replies, pointing out that the Council did not identify itself with the statements of individual members during a debate.[27] Yet it could scarcely dissociate itself from its academic head. It was Hulme who had been quoted as saying that the undertaking might cost £500,000. Was there a private local architect who could put his hand on his heart and say he could complete plans, costing say £40,000, in two years? The Government Architect had conceived the overall plan and it was logical that he should begin its development.[28]

That, however, was merely a storm in a teacup, to be forgotten in the fullness of time. What followed was a dispute, precipitated by the Rector, which divided the Council and distanced it from the School, the profession and the public. The University's centennial history relates how Hulme declined in the Senate in 1949 to vote for rescission of a motion that a School of Forestry, if and when it was established, should be at Auckland.[29] Canterbury was at first incredulous and then furious; but Hulme appeared to have learned nothing from that affront to parochial sentiment and pride. With a fine disregard for everything Canterbury held dear he acquiesced early in 1952 in what seemed to be a Machiavellian plot to cut the Engineering School to ribbons.

Hulme and the Council chairman, W.J. Cartwright, went to Wellington and were told by Beeby he could not recommend the proposal for a School for 430 students to the Government. He argued that the aim of an eventual 950 engineers per million of population required an output of 120 graduates a year, but perhaps this should be extended to 140.

There should be 25 graduates a year for every 100 students so about 560 would be needed to produce a target of 140. Add a provision for postgraduate students of say 40 and the total number of students would be about 600. Then — and here was the real rub — divide the total by two, giving Canterbury a mere 300 students, a miserable 20 more than it already had.[30]

The mathematics might have been acceptable, but the premise was not; yet Hulme welcomed the proposal when it came before the Council in February 1952 as "excellent and very reasonable", though he did not disclose the prior discussions with Beeby. An equal division would, he thought, be simpler "as equal support could be given and there should be no quarrels about advantages." And, he added with a trace of the priggishness that prejudiced his support at Canterbury, "I happen to know that if we go ahead now, things are likely to proceed satisfactorily."[31] This hint from the corridors of power did not sway H.F. Baird, chairman of the Council's Engineering Committee. If a bird in the hand was worth two in the bush, said Baird, this one might leave a sticky mess if seized too quickly. C.E. Fenwick, a consulting engineer, moved that the matter be referred to the Engineering Sub-committee and asked that the views of the College's Engineering Advisory Board (a group of leading Christchurch engineers) and the Institution of Engineers be heard; but his amendment was lost after Hulme observed that "this is purely a policy matter requiring common sense rather than the expert knowledge of engineers."[32] The Council went along with the proposal, little realising it was lighting the fuse to a barrel of gunpowder. If the School had any doubts about the esteem in which it was held, the subsequent reverberations must have seemed as encouraging as they were damaging to the College Council. As the *Press* observed:

> Cambridge is famous for its Cavendish Laboratory, Edinburgh for its medical school, Oxford for the liberal arts. Throughout the world Canterbury College has been known as the home of New Zealand's School of Engineering.
>
> When it becomes known that Canterbury has readily agreed to hold up the progress of the School to admit the parity of Auckland, much of the prestige of the College will be lost . . . in effect the Council acquiesced in the proposition that the Canterbury School should cease

to grow while the Auckland school trebles in size . . . Canterbury is in danger of falling to subordinate status.[33]

The evening paper said the Council, after the briefest of consideration, had thrown away the one thing that made Canterbury famous throughout the world — the National School of Engineering.[34] Letters of protest from individuals and organisations flowed in to the Registrar and the Canterbury branch of the Institution minced no words. There had been material evidence in recent years (it said) that the Council was failing to appreciate or accept the public responsibility inherent in its control of the School. The decision to limit the growth of the School was "a serious error of judgement which should be corrected without delay."[35]

The letters were easier to answer than the protests from the School. Hopkins, who had made no secret of the fact that he came to Canterbury because of the prospects of a new School,[36] led the attack by criticising Beeby's numbers game and pointing out that direction of students would be required to implement the suggested allocation. That direction would need to be endorsed ("aye, there's the rub") by the College Council.[37] The staff of the School said they were appalled at the Council's failure to consult both the Engineering Committee and the School. In a letter to the Rector they claimed the Council must seriously have underestimated the value of the School or it would not have circumscribed its future development.[38] The 21 signatories of the letter by what was then called the sub-professorial staff were met by Council members, but were not swayed. "We regard the Council's action as being unjustifiably precipitate and at variance with the accepted traditions of university administration," they said in a further letter.[39]

The correspondence columns of the newspapers were dominated by the topic. Indeed the sound and fury reached such a pitch that Cartwright felt impelled to respond; but his explanation only aggravated the dispute. After claiming the Council had not taken hasty action (though "it might appear that it took rapid action") he let slip the fact that he and Hulme had prior discussions with Beeby.[40] Baird was furious that his committee had not been consulted and thought the absence of discussion by the Council "arose from shock following severe starvation of information."[41] The Dean (Rastrick) issued a statement expressing surprise at the Wellington meeting and MacElwee said that if the decision was taken under pressure from Beeby "the traditional freedom of action of university governing bodies is seriously threatened."[42]

Clearly the College was riven; but a joint meeting of the Engineering and Policy and Development Committees failed to heal the breach. Rastrick accused the Council of having no confidence in professional engineers. Hopkins said his concern was not with ratios but with an adequate School which could expand if enrolments increased. No university had ever been planned for sufficient numbers, he added prophetically; to reduce the plan (for the School) "is not only illusory but ludicrous." But it was a subsequent exchange at that meeting which illustrated the problem. Hulme's convictions were not trimmed by parochial considerations and in acting for what he saw as the national good he trod clumsily over local aspirations. The School had pride in its past and confidence in its future but the Rector, with an exasperating intellectual primness bordering on arrogance, told the engineers at the meeting that they should aim higher.

> MacElwee: The question at issue is the number of engineers required.
> Hulme: We should advance our points from a national point of view.
> Hopkins: No, we should advance our points from a Canterbury point of view.
> Hulme: We cannot expect to put up to a national body a purely Canterbury viewpoint.[43]

That meeting was inconclusive and the furore continued. Beeby was questioned by the press and there was a special article by an unnamed "engineering graduate" (by popular report, a staff member, Arthur Lush) which likened the College Council to a certain Mrs Partington, who earned immortal ridicule during a great storm on the eastern seaboard of the United States by taking a broom to sweep back the water coming under her door, oblivious to the fact that she was trying to sweep back the Atlantic. And that was only the start. The article finished by asking:

> In the present case do we see, thinly veiled, the possible emergence of an educational dictatorship which scorns advice from one faculty (eventually it may scorn them all) and values power and authority above democratic rule and the interests of education?[44]

The pressure could not be denied and at the end of March 1952 the College Council duly rescinded the decision and agreed that the size and capacity of the School should be reviewed by the Engineering and

Policy and Development Committees. The debate was not particularly acrimonious, but during it the first public criticism of the Rector was heard. Baird accused him of indiscretion and haste. The Rector responded by accusing Baird of putting over "more guff and emotion than I've heard in a long time . . . He has said no more than 'Go for it boys and get all you can.'" When Baird proposed the establishment of a committee to press the case for the School, the Rector's name was, significantly, not included; and Professor H.N. Parton (Chemistry) noted the omission. If Baird had no confidence in the Rector in the negotiations, he should say so, Parton said. "This way of doing it is deplorable."[45]

The newspapers expressed general satisfaction, though the *Press* noted that the Rector's speech was virtually in opposition even though the vote was unanimous. More weight, it said, should be given to the opinion of Ward that the Engineering School and its students were the backbone of Canterbury College.[46] The outcome was an invitation to Beeby to meet a special Council committee (on which the Rector had a place) because a reduction in the size of the School "might be injurious to engineering education in general and Canterbury University College in particular".[47] The meeting was held but no settlement was reached. Beeby reacted by wickedly suggesting the matter be referred to the U.N.Z. Senate, where it was feared (with some justification) that it might languish for years.[48] That idea, said Baird, was an outcome of approaching Beeby in the first place. The approach should have been through the local member of Parliament,[49] who happened to be the Prime Minister.

A compromise was needed and it came from a number of sources. The Engineering Advisory Panel was cautious about predictions, but thought the methods already used could give "a sufficiently reliable indication." Applying them, the panel concluded that a New Zealand-wide undergraduate roll of 628 engineering students, excepting chemical engineering, could be accepted as a conservative estimate for 1975.*

Canterbury's share of this total "may be expected to be not less than 424, to which should be aded 30 postgraduate students, giving a total roll of 454".[50] (The roll was actually 841 in 1975).

The estimates carried some weight, but not as much as Beeby, who clearly meant business. If the building was delayed, he said pointedly,

---

*It was conservative. The Canterbury and Auckland rolls in 1975, excluding chemical engineering students, were more than twice that figure

there were plenty of other school buildings needed. In any case it was just as bad to over-estimate for a building as to under-estimate.[51]

The point was difficult to ignore and the special committee to meet Beeby* had its back to the wall. Finally, and reluctantly, it recommended to the Council that a figure of 300 be used as a basis for planning; and in June 1952, the Council returned, with only three dissentients, to that figure after the Chancellor of the University (Sir David Smith) warned that the matter would indeed go to the Senate if agreement was not reached.[52] The desire to get something started overcame any reservations. Argument about future size could wait. Even Baird accepted the motion, though he commented that Beeby was a State servant and not yet "Director of University Education". The Council had, he thought, acted "gauchely" in its dealings with Beeby.[53]

Public reaction to yet another twist in Council policy was remarkably subdued, the *Press* merely commenting that the Council's original mistake was in allowing itself to care little for the prestige of the School. After it had agreed to equality of numbers with Auckland there was little it could do to retrieve the situation.[54] Only time would tell, said the *Star-Sun* whether the decision was correct or merely a concession to expediency, but at least the matter had been threshed out and some uncertainties removed.[55] Subsequently Hulme and Rastrick visited Wellington for talks ("unofficial, of course") and the first result was a Council decision to accept a net area of 90,000 (107,000 gross) square feet for the School.[56]

The engineers had not exactly won the battle, but neither was it lost. A new School could now be planned and perhaps expanded as required. Yet there were scars. The staff had been anxious to improve the School's prestige and to continue to attract students from all parts of the country. They feared that if 600 students were to be divided between the two schools, Canterbury would be full immediately and future students would be directed north. They also feared that once a new School was built, there would be no extensions until the rest of the University had been established at Riccarton. Building costs were rising and an adequate School initially would be as cheap as extensions later. But hard-pressed departments on the city site were not particularly sympathetic. Accommodation was under pressure in every faculty. By 1950 total enrolments stood at 2583, the highest ever. Courses began

*It consisted of D.W. Bain and C.H. Perkins, both future Chancellors, Cartwright, Baird, Rastrick and Hulme.

to multiply and new chairs were approved. And the College looked hungrily at the space the engineers would vacate. Near the end of 1952 the Council virtually demanded additional accommodation for 1953, defying a Cabinet ruling that no new property or buildings be authorised for the College pending completion of the School.

Completion? There was no sign of a start. Hulme told the Council: "We may not get to Riccarton for five or seven years and during that period students of university age will increase 30 or 40 per cent."[57] On the cramped city site repeat lectures were required and evening laboratory classes were proposed; but the staff could not be worked round the clock and additional staff would require more accommodation anyway. As it was, eight lecturers had no office and 32 worked four to a room.[58] Indeed a report to the Council described some lecturers' rooms as dogboxes.[59] Diogenes may have enjoyed his tub, but in the mid-twentieth century even the most unworldly of lecturers felt entitled to an office, a desk and space for books.

Repeated letters to the Minister (R.M. Algie) brought only the dustiest of replies and a deputation waited on him in Wellington. Algie was prepared to consider relaxing the ban on city site building if asked by the Senate's U.G.C., but as for the Engineering School he could not put that matter before the Cabinet until he could see more closely what finance was involved.[60] His hand was jogged early in 1953 when fire gutted the east block of the College in Worcester Street. If there had been any notion of abandoning the move to Riccarton at that stage, it went up with the smoke. Geology staff and students became the first academic colonists of distant Ilam where, if opponents of the move were to be believed, moas (in the words of one Public Orator) might be seen tramping westward at dusk. When the Geology Department returned to the city site after the East block was restored, the School of Fine Arts moved to Ilam into the Okeover homestead and temporary studios; these were also destined to go up in flames in 1975. The Library promptly spread into the space the school vacated. Other departments moved into houses bought or rented about the city site. The first round of departmental musical chairs had begun, and Ilam was inextricably part of it. Asked early in 1953 to sum up the College's building needs, Logie replied: "To go to Riccarton".[61] And the priorities were a School of Engineering followed by a block for chemistry or physics, conversion of a house for a temporary Students' Union and the preparation of playing fields.[62]

Algie had some justification for not committing himself in 1953. The geographer, Professor George Jobberns, on his return from study leave in North America, became the spearhead of a campaign to rescind the 1949 decision to go to Ilam and instead to develop the area west of Montreal Street as an educational precinct. Many staff, exasperated by Government intransigence, supported him. So, too, did the *Press.* But the Council, under fire from leading articles, sheafs of correspondence, a well-attended public meeting, staff meetings and a deputation, remained firm. Again and again it pressed Algie. The Chancellor and Vice-Chancellor of the University waited on him and the Institution of Engineers urged him to give priority to building new schools at both Canterbury and Auckland.[63] The tone of Logie's letters began to change.

> The College Council has directed me to ask if you would be kind enough to inform them whether or not Cabinet has yet reached a decision about the plans for the new Engineering School at Riccarton. If no decision has been reached the Council would be grateful if you would inform them of the probable date of such a decision.[64]

The silence was unbroken, leading the *Star-Sun* to wonder, in an editorial headed "College of Dreams", whether the transfer to Riccarton within the measurable future would become "just another lost cause."[65] And since the *Press* perceived fewer difficulties about expansion on the city site it called for abandonment of the "ambitious scheme" for Riccarton.[66] Algie declined to meet a Council deputation; a protest meeting of local M.Ps. resulted in a sheaf of apologies and the attendance of only one member, H.R. Lake; and Algie sprang a further delay by asking Canterbury and Auckland to discuss possible common plans for the two schools to reduce architects' fees.[67] Representatives of the two colleges met, but what emerged was a feeling of hopelessness rather than any common plan. Even a grant or loan from the City Council was contemplated.

The campaign by Jobberns had in the meantime reached something of a climax, but a sharp debate in Parliament served both to clear the air and to put the first skids under opponents of the move. Canterbury M.Ps. clearly believed that if Christchurch continued to quarrel about the site there would never be new buildings. Some Opposition speakers claimed the Government had seized that argument as an excuse for doing nothing. But J.K. McAlpine, a member of the College Council (1950-59) chided those who criticised it. The College was designed to accommodate 600 students and was now trying to accommodate 2240.

> The College is deplorably overcrowded (he continued). There are sheds, shacks and shanties clustered around the College, even on the sacred quad.[68]

The year ended without a positive signal from Algie, but the controversy in Christchurch seemed to have run its course and the *Star-Sun* rebuked its morning counterpart:

> An idea, born late and pursued with more wishful imagination than appreciation of plain, unpalatable facts, resulted in a dream, unlikely ever to become substantial, of a university rising where there is no room for it. The idea has caused hardly a ripple of real public interest. Indeed most people probably promptly dismissed it as yet another example of how fertile Christchurch is in talking round a subject instead of getting on with the job.[69]

Lush, in a special article, contrasted the dispute over siting Auckland's buildings ("we have eyes but no site") with that in Christchurch, where some citizens "with speech but no vision will not see that you cannot put 10 gallons into a pint pot, no matter how old and beautiful the bottle."[70] Finally the Council scotched any suggestion of wavering about Ilam and the reaffirmation of its policy was timely. But other problems caused Canterbury's mana and morale to fall to a low ebb in 1954. Hulme, under pressure, resigned. Shortly afterwards his daughter was arrested on a charge of murder and the subsequent trial was reported world-wide. The only lightening of the general gloom was Algie's announcement that the Government accepted the progressive transfer of the College from the city and agreed that the first building on the new site would be the School of Engineering.[71] That was followed promptly by the long-awaited announcement that the Government Architect had been authorised to prepare preliminary sketch plans for a building with a ceiling price of £450,000.[72] Three months later preliminary drawings were approved by the Cabinet, "the first positive step (Algie said) "in the long hoped for building".[73] The drawings were not accepted uncritically and a planning sub-committee* had grave doubts about some features. What, for instance, was the artistic relevance of the "protuberance" of the circular lecture theatre? It was the first but not the last criticism of the lecture theatre, which for a brief period after its completion was symbolic of the new University.**

The sub-committee disliked the "sameness" of the height of the core block and questioned the proposed finish.[75] The Government Architect, F.G. Wilson, placed amended plans before the Council early in 1955 and said he hoped to call tenders in a year and see completion in 1959.[76] The Council applauded this resolution, but the sailing was not to be as plain as that.

The engineering professors, not unnaturally, subjected the plans to detailed examination. They were aware, no doubt, that many eyes were now focussed on the endeavour and were determined that no errors should mar the final plan. They sought a host of minor changes and some not so small, including the allocation of space saved by reducing corridors to the core block. It was messy and time-consuming and when Dr F.J. (later Sir John) Llewellyn — the iron Vice-Chancellor[77] — replaced Hulme he took firm control, and demanded that all requests for changes be made through him. A grasp of detail and a new authority are revealed in the correspondence which flowed from him to professors and architects alike. He was concerned, for instance, that the large windows facing north would pose problems*** and he commented that the library layout

---

*It consisted of Bain, now Council chairman, Ward, Cartwright, MacElwee, Professor Edward Percival, Perkins and the Rt.Rev.A.K. Warren, who was also to become Chancellor.

**Standing on a stalk with few other visible means of support, the round, domed lecture theatre soon became known as the Mushroom and even a Prime Minister once became lyrical about it. The inverted, saucer-shaped dome would, said Keith Holyoake, span more than 50ft but would be only about three inches thick, the relative thickness of an eggshell. It would be, he thought, perhaps the most interesting and attractive feature of the School.[74] The University history (p348) says it accommodated 180 of what, perhaps maliciously, were said to be the most expensively seated students in the Southern Hemisphere. Students liked the design, but they said it prevented lecturers from being cornered.

***They did initially. MacElwee claimed that sunlight streaming through them would make it difficult to read dials on instruments in the laboratories. He was told to take "remedial" action:placing a book in such a position as to shade the scale. But the architect did acknowledge that the windows posed a problem, especially when it was reported that temperatures of up to 110 deg.F. had been recorded in the professor's study. Subsequently W.D.H. Smith, who had been appointed Assistant Registrar (Buildings) in 1961, was called upon to have external "eyebrow" shades fitted and the professors' studies cooled.

appeared to make provision for almost every contingency but the storage of books.[78]

This attention to detail paid dividends. As another notable instance: the lecture rooms were cramped but in the circulation area outside there seemed to be an excessive number of structural pillars. Would it not be possible, the Rector asked, to move the walls a few feet into the circulation area to take in the pillars?[79] Yes, was the reply; but it meant encroachment on the courtyard. So be it, was the response. Four lecture rooms had been planned along the concourse. Llewellyn proposed three larger rooms and gained a fourth by moving cloakrooms and lavatories elsewhere. Slowly the plans were adapted to enable the School to accommodate up to 400 students.

These changes took time, and 1956 came and went. The architect reported that it was a complex job already involving 40 sheets of architectural drawings.[80] In all, 140 sheets would be required. The Council was displeased and Taylor was moved to say it was a sorry day when planning was entrusted to a Government department,[81] a sentiment echoed by Hopkins, ever prepared for a bold approach, who said the architect's staff should be working at Riccarton. "Unless he is prepared to adopt this course it will be quicker for us to terminate his appointment and engage a local architect to complete the job."[82] That advice was not heeded, but Rastrick was commissioned to design the mechanical, electrical and other services.

Algie assured Parliament that the School had high priority[83] but the Rector went higher. Would it not be possible, he asked the Prime Minister, to hasten the completion of plans and the calling of tenders? He outlined all the reasons and then gave another:

> It had been my intention to invite you to lay the foundation stone of the School in late August 1957 when the contractor had commenced building . . . doubtless you will agree that it is essential building is actually under way before the foundation stone is laid. The recent history of several foundation stones in and around this city which stand in splendid isolation and loneliness warns me not to proceed with such a function until the contract is well on its way.[84]

Holland was too shrewd to rise to that bait; in any case, another snag delayed progress. Gas heating had been planned for the School pending coal-fired steam boilers when other buildings arose at Ilam. But the

Gas Company was unable to provide an economic supply and it was decided at a late date to proceed with the first stage of the boiler-house to heat the School, a step which caused the *Press* to comment that the Ministry of Works seemed to have made a hash of heating the buildings.[85] It also alerted the neighbours, who established the Ilam Action Committee to ensure their homes were not in future to be shrouded in smoke and fumes.

As the size of the School increased, the cost crept up. In mid-1957 the Commissioner of Works asked Beeby to seek Cabinet Works Committee approval for a School of 126,000 square feet, costing £600,000. Three years before an estimate based on 110,000 square feet was £505,000. There were a number of "extras, including the first stage of the boilerhouse, an underground tank and concrete water tower designed by Hopkins* and an additional 16,000 square feet in the School itself. There had also been a 5 per cent rise in building costs since 1954.[54] The Council gave final approval to the working drawings but the proposed date for calling tenders came and went. Anxious soundings were made in Wellington. The professors prepared press statements, Hopkins characteristically giving his the heading, "Impromptu statement for the Press," and finally, in mid-August, Holland announced Cabinet approval. Five weeks later the tender of John Calder Ltd for £614,972 was accepted and authority was given also for purchasing steel, pumps, cranes, hoists and pipes as well as groundworks, bringing the total to £681,292. Of even greater interest to Llewellyn,who was about to assume the title of Vice-Chancellor, was the timing. Thirty months were allowed. If the target were met the School would be ready for the start of the 1961 academic year.

The *Press* seemed at last to have dropped its campaign against the move. The University, it said (properly; the College was named "University of Canterbury" in 1957 even though autonomy had to await the passage of the University of Canterbury Act 1961) "may be congratulated on the start of the first permanent buildings on the new site and should not be considered greedy if it immediately asks for more."[88] The Council

---

*Hopkins wanted to conceal the pipes carrying water to the tower and recalled a Father Brown detective story in which a thief was not suspected because he was in uniform and had a right to be where the theft occurred. So Hopkins designed the tower with three concentric concrete pipes which both carried the water and supported the tank.[86] The high water table posed difficulties for installation of the tank and at one stage two members of the Canterbury Underwater Club were called in to work 20ft under water anchoring a sump and directing the pouring of a 4ft thick slab of concrete.

certainly heeded that advice, approving new equipment for the School costing £226,000 as well as space schedules for the new science buildings.[89]

What would it look like, this key building in the University's ambitious plan? The general outline was well enough known after seven years, but what of the particular? The exterior finish, one report said, would be one of the most unusual in New Zealand. The concrete blocks would be poured on to a facing carpet of coloured sand, gravel and chips, the specifications requiring basalts from Halswell and McCormack's Bay, Chertsey argillite, yellow, grey and blue limestone from Waiau, cream and red limestone from Hanmer Springs, an alabaster quartz from Duntroon, white marble chips and various coloured sands . . . "the masonry is expected to give the effect of brightly-coloured iridescent crystals".[90]

There was precious little of that vision to be seen when the foundation stone was laid, but good progress was made initially with the contract. Delays, however, seemed inseparable from it. A nuclear engineering laboratory with special plumbing and drainage had to be hurriedly designed when a sub-critical nuclear reactor was presented to the University by the United States under President Eisenhower's "atoms for peace" programme. The reactor was removed after nearly 20 years' use when it became clear New Zealand would not contemplate nuclear power generation this century. Hopkins could not specify the strengthening required for Civil Engineering laboratory floors until he knew what equipment he would get, and that awaited a special grant. Minor alterations, also the result of close staff oversight, slowed progress. The Council inspected the work in November 1959 and announced that the building would be progressively occupied during 1960. It was half a loaf, perhaps, but much better than no bread. As engineering staff vacated their offices on the city site maintenance staff moved in to create Room D, the largest lecture room in the University. Accommodating 286 students, the room was of constant concern as a fire hazard, but it served effectively the need to reduce the repetition of lectures. Meanwhile lecture rooms at Ilam were pushed ahead to be ready for the second term, though laboratory and drawing classes still had to be taken in the city and a shuttle service of red buses was laid on at University expense. It was a messy solution and it was made worse by delays in the delivery of furniture and equipment.

Earlier, the staff had estimated that the cost of new equipment would be more than a quarter of a million pounds and removing and reinstating equipment would amount to £22,000, with an additional £40,000 for workshop equipment. There was also a 15-page schedule of furniture and fittings. The Cabinet allocated £140,000 for equipment but Hopkins, who was on study leave, was furious at the way it was divided, especially the allocation for workshops. It was difficult to express, he told Llewellyn "the deep sense of frustration, if not betrayal, I feel."[91] Who, the Rector responded, was betraying whom?[92] Strong words, came the reply, were an indication of strong feelings.[93] Llewellyn took the problem to the Council which endorsed an agreement among the professors that the Civil, Electrical and Chemical Departments equip departmental workshops from £20,000 allocated for the purpose with the larger central workshop in Mechanical Engineering receiving the balance.[94]

Elsewhere there was a united front. The professors (the Rector wrote to Beeby) believed that because carpet was provided for the heads of secondary schools, they deserved similar treatment[95]. Their suites got fitted carpet. There was no heating and in the winter term 24 electric radiators were purchased. Batteries of them took the chill off lecture rooms in the early morning before going to staff rooms for the rest of the day.

If this early part-occupation was something of a bonus for the University, there was anxiety about completion of laboratories and drawing offices by the start of 1961. The official opening of the School in October — the first new building for Canterbury for over a quarter of a century — could not disguise an unfinished look. Llewellyn told the architect that progress, especially in electrical work, was far from satisfactory. Discrepancies between plan and finished work were continually being found. The contractor was annoyed at suggestions that he was dilatory and blamed "excessive changes and alterations,"[96] a claim that was given colour by a complaint of departmental delays in returning quotations for and answers to queries about furniture and equipment orders and specifications.[97] If the staff had to accept delay, clearly some of it was of their own making; but the slow installation of cranes was also a factor. Hopkins made a close inspection in the summer vacation and alerted Llewellyn. Only "a rocket of unprecedented magnitude" would, he said, get action.[98] In the end there was an untidy finish, but as autumn turned to winter in 1961 the engineers were in full occupation. Even "Lake Calder", an untidy pond that lapped foundations after heavy rain, disappeared.

The roll that year was 513 and a projection made then forecast 657 by 1967. (It was actually 677.) Changes during the protracted planning of the School and its construction meant that it could accommodate half as many again as the 300 for which it had been designed a decade before. But it was still too small, a failing emphasised by a student prankster at the opening ceremony. "School of Engineering", said his neatly-lettered sign at the main entrance, adding; "Scale Model". Three separate cases were made for more space from 1959 on, but initially the Senate Grants Committee was not impressed. In 1960 it commissioned two U.N.Z. officers to predict New Zealand's engineering manpower requirements a decade hence. They forecast that between 1200 and 1400 engineers would be required per million of population. That meant that between 1000 and 1200 engineering students would need to be undergoing training in the universities 10 years hence. Therefore, schools of 550 at Canterbury and Auckland "should be the present objective . . . for the forseeable future."[99] Llewellyn was convinced that the figures were too low and insisted on provision for 700 students who would require extensions of 55,700 square feet, including a new library, more lecture rooms, bigger laboratories and a new block for Chemical Engineering. Before prosecuting the case, he was appointed Chairman of the new University Grants Committee which henceforth would consider and investigate all university requests. But Canterbury seemed not to appreciate how much the old order had changed. The new Vice-Chancellor, L.L. Pownall* arranged for a deputation to wait on the Minister of Education (W.B. Tennent) with a 35-page submission for extension of the School, at the same time explaining to his predecessor that the decision "in no way implies a lack of confidence by my Council in you as Chairman of the University Grants Committee."[100] No reply is on file, but one can imagine Llewellyn's face darkening with displeasure. Perhaps that was why the deputation was bowled an impossible ball. Would Canterbury give the extensions top priority? There could be no clear answer to that. Completion of the Science buildings at Ilam was essential, as were other works.

But the climate was different in the sixties. The Government was by then aware of the magnitude of buildings required for higher education and of the backlash that reducing student intakes might provoke. Big

*Pownall, Leslie Leigh (1921-);Educ Palm.Nth BHS, Wgtn Trg Coll., VUC & CUC (MA.),Ph.D.;asst CBHS,41-46, lect ChCh Tchs'Coll.46-47; Whitbeck Fellow, U Wisconsin 47-48, lect Ardmore Tchs' Coll. 48-49, lect.AUC 49-51, lect CUC 51-60, prof.Geog 60-61, VC 61-66; Clerk of Senate, London, chmn NSW Env.Com.

might not always be beautiful, but small was certainly unwise politically. In addition, predictions were made with rather more accuracy than a decade before and political responsibility demanded that greater provision be made for future needs. So it was that the U.G.C. in 1962 gained general approval for its proposition that New Zealand's needs for engineers required places for 1400 engineering students. A new school accommodating 700 would be built in Auckland and the Canterbury School extended to take a similar number. Canterbury submitted area schedules and sought an additional 58,690 square feet, but Llewellyn had no intention of tediously negotiating room by room, laboratory by laboratory. He would, he told Pownall, be prepared to recommend extensions to bring the overall gross area up to 200,000 square feet.[101] Pownall agreed, after hasty meetings with the engineering professors and the Executive Committee of the Council, to accept the proposal with only one caveat: he asked that the final area for the whole building be left in abeyance until sketch plans were drawn since it was conceivable that a figure somewhat exceeding 200,000 square feet might be the most sensible design.[102]

At the end of 1962 the Government approved a quinquennial national university building programme. For Canterbury it included science buildings, the first two stages of which were already under way with others at tender; a Students' Union, for which a subsidy had been approved; the Library-Arts building and other buildings for the Arts faculty; a Registry; and last, but certainly not least, the Engineering School extensions. It was, said Pownall, a milestone announcement.[103] The extensions totalling 48,967 square feet and costing $1,390,200 were completed by five different contractors between 1964 and 1967, the most significant being the new library, and the five-storey Chemical Engineering block. The total cost of the School,though inflation makes the figure difficult to appreciate, was $3,109,162. The only major addition since was an extension to the library, partly for computer terminals, completed in 1984.

Engineers with any pretensions to aesthetic appreciation found it difficult to express pleasure about the appearance of the buildings. Back in 1951, Harman hoped the design would not too readily favour the type of building alleged to be modern because "the modern of today can look very out of date the day after tomorrow." Harman favoured a Georgian style.[104] The new School was neither. When he announced the letting of the contract in 1957, the Prime Minister (Holyoake) had been moved to say that every effort would be made to preserve the atmosphere of University buildings while the architecture of the University of Canterbury would be continued in the use of local Halswell stone.[105] Alas, though Halswell stone was used to cover the ground floor wall panels between the columns of the concrete frame structure, that was the principal architectural similarity and one looked in vain at Ilam for evidence of the sure hand of a Mountfort or Hurst Seager. John Oakley, a lecturer in fine arts, caused annoyance and embarrassment by referring publicly to the long, squat buildings apologetically facing Creyke Road, the windows covered with brown paper and aluminium foil to keep out glare and heat. When the box-like lecture rooms were added below the circular lecture theatre "any aesthetic excuse for the colossal expense of building it on stilts becomes pointless."[106]

But functionally the buildings were satisfactory, the space was a joy and the design served the School well. And there were pleasing features. The spacious concourse was quickly appropriated for several major University receptions and in the courtyard garden, a faint reminder of the old quadrangle, students ate their sandwiches on fine days and a pair of ducks raised an annual brood. Time, landscaping and the growth of trees and shrubs have, as with the rest of the University, softened the lines, smoothed the angularities and enhanced the appearance of the School. The buildings belong. Subsequent buildings on the campus were a distinct improvement on the first; but the School proved serviceable, comfortable and adaptable after initial faults had been repaired. And for those who cared to recall the Auckland-Canterbury controversy, there was one satisfying footnote. Professor N.A. Mowbray, Dean of the Auckland school, was reported in 1963 as saying that his school's roll had dropped because prospective students were attracted by the new buildings at Canterbury.[107]

H.J. Hopkins

L. Kay

J.R. Burton

S.R. Siemon

A.M. Kennedy

CHAPTER TWO

A Library of Substance

Few specialist libraries would have rivalled the Engineering Library when the new building opened in 1967. It was in many ways a trial run for the Main Library completed at Ilam seven years later; it gave users some inkling of how library design, furniture, equipment and services had advanced as the knowledge explosion of the sixties spread; and, in contrast to the first engineering buildings at Ilam, it benefited from later Government recognition of the space and facilities needed for higher education. But its development was far from smooth. Indeed its fortunes veered so much initially that its eminence now seems almost as much a matter of luck as of sound planning and persuasive supporters.

On the city site the Engineering Library had been an orphan— "this departmental Cinderella" was the description given by Lyall Holmes as late as 1945—and the legacy of Scott and his three successors weighed heavily on it. They reminded C.W. Collins,* Canterbury's first qualified librarian, that it was their library rather than his** and Holmes's complaint that no one was responsible for its life or death was justified.

But when, at the end of 1945, the care of the library was made the responsibility of a full-time librarian, the engineering staff began to realise there was more to being a librarian than stamping books (after ensuring, as one instruction to the attendant put it, that the date stamp was changed daily) and keeping the shelves tidy. Both Helen Holmes (nee Jobberns) and Sheilagh Burke demonstrated the value of

*Collins, Clifford Wallace, 1909-1979: B.ChCh, educ CBHS, CC(MA), Michigan U (ABLS), apptd Librarian 1934, ret. 1971. He was responsible for classifying, cataloguing and shelf-listing the complete holdings of the Library, which grew from 15,000 to 300,000 volumes during his term.

**"I find new books in the Library that I have not seen," Powell, then Dean, wrote to Collins in 1943. "I shall be glad if you will revert to the practice of submitting all new books ordered by me to me for inspection and classification prior to their being placed on the library shelves."[1]

an effective library service for the School, producing an informative and chatty version of the *Library Record* to keep both staff and students informed about developments. Collins had always foreseen the Engineering Library as providing an information and reference service once the fundamental bibliographical and library work was up to date. He hoped the appointment in 1948 of D.L. Jenkins, a man with a background in science as well as librarianship, might give it that new dimension; but planning the new library took precedence after the decision a year later to move the School to Ilam.

Many academics no doubt looked on Collins as the quintessence of the conservative, dry-as-dust librarian; but his meticulous attention to detail — some would say fussiness - obscured a total devotion to the well-being of the University and its Library. His concern for the future of the Library in particular is manifest in hundreds of notes and memoranda in his tiny, crabbed handwriting contained in the Library's files; and nowhere was that concern expressed more forcibly than in a memorandum he and Jenkins prepared warning against the folly of using the old Engineering Library as any sort of yardstick for the space and facilities that would be required for a new library. In what must have seemed almost like a quantum leap, they proposed that it should be of 7100 square feet for up to 100 readers, and up to 8400 square feet for 195 readers.[2] They assumed, correctly, that much greater use would be made of the new library, partly because changes in teaching and learning demanded more intensive use of books, periodicals and abstracts and partly because (as Parkinson would subsequently rule) a better library would of itself attract more use. They also suggested that once out in the Ilam wilderness, students would stay on the site all day and would perforce study in the library.

But the Engineering Library did not at that stage have the academic advocates nor the resources it could confidently expect a few years later. If Collins's warning was noted initially, the report was given less attention when specifications were drawn up. Instead, a curious exercise was undertaken to determine how many of the 250 × 1.6 = 400 + 30 postgraduates = 430 students would use the Engineering Library at any hour of the day. The answer ranged from 160 to 270 and "it is reasonable to assume" (as the report put it) "that provision should be made for 110 readers."[3] At 30 square feet each they took up a good half of the space sought. A book stack of 14,000 volumes was assumed, requiring 1270 square feet; the circulation desk, catalogue, periodicals display and offices were allocated 900 square feet; and a total of 6070 square feet was sought. That compares with the 1319 square metre (14,198 square feet) of the library today.

The Rector was not impressed. The College should not adopt the view that the initial submissions should be excessive and any excess bargained away subsequently. On the contrary, said Hulme, specifications which could be justified should be put up. This need not involve cheese-paring.[4]

There was, however, some paring of the Library as planning of the School proceeded. In one plan it ascended from ground floor to first floor, somewhat to the surprise of Collins. There is a tendency, he said in an "unofficial" letter to the architect, to refer matters to the heads of the four teaching departments, and, unintentionally, to forget the Librarian. "This has caused difficulty in the past and with the best will in the world is likely to recur. Do not take it for granted that Library opinion has been included in all documents prepared."[5] Collins was not strongly opposed to a first-floor library — he saw an advantage in extra quietness — but he preferred the ground floor, partly because of his experience with the College Library. It was only the provision of a basement there that had prevented complete chaos and he looked forward to gaining extra space under the new library for future expansion. "Though the use of partly excavated space is apparently being frowned upon (heaven knows why!) I continue to recommend provision for later development of a basement" Collins wrote.[6] The engineers, if not heaven, could have told him why "partly excavated space" was frowned on. The water table at Ilam, as those who undertook excavations for various purposes were to find, was uncomfortably high and an underground library might also have become an underwater one.

The library was finally located on the ground floor overlooking the central courtyard, but then it suffered severe pruning. As planning for the School proceeded, it became abundantly clear that more than 400 young men could not be sent into the distant fields of Ilam and expected to survive a long day without food or drink. For many years there would be no Student Union on the new site to solve this pressing problem. Accordingly, the plans were searched for an appropriate site for a "temporary" student cafeteria; and one was found by slicing a portion from the space allocated for the library, which was left with a mere 3150 square feet.

"Temporary" is capable of a variety of definitions, but the term is perhaps most loosely defined within the university system. After all, Canterbury's first School of Forestry, closed "temporarily" in 1934, did not reopen until 1970. In due course the library occupied space which is now the Senior Common Room and the "temporary" cafeteria was built alongside it. One was stacked with 6000 books and periodicals from the city site in mid-1960 and thereafter acquired books at the rate of one a day. The other was restocked with pies, buns and drinks daily and at times was far busier. Indeed, it was soon indispensible. When extension of the School was discussed that year, the Dean (Siemon) told Llewellyn it could be assumed the cafeteria would be converted to library use, but if not the only possible solution would be a new building, with the existing library being used merely as reading rooms.[7] As plans were prepared for the first stage of the Student Union, there was growing student reluctance to face the inconvenience of traipsing in all weathers to the far side of the campus for sustenance. Subsequently Ensoc asked formally for retention of the cafeteria and the request was, fortunately, granted, although the cafeteria was trimmed to the proportions of a snack bar when meals became available at the Student Union.

It was fortunate because it precipitated the idea of an entirely new building at a time when library horizons were suddenly widening, when planning for the Main Library at Ilam was beginning and when the inadequacies of the first library were becoming all too clear. Barbara Hale, who resigned from the library in 1960 to become Deputy Librarian at the University of Western Australia, reported that she could shelve only 5000 volumes, some 25 per cent of the total stock. The remainder were stored in 30 nooks and crannies — one accessible only by a manhole — which Collins found on the city site, and in the Ilam homestead. It was almost impossible to display the 250 periodicals to which the library subscribed. Of 76 chairs, only 67 were at tables, most of which had nine readers rather than the six for which they were designed.[8] The library itself was hot and noisy and it was difficult to concentrate therein, according to the Dean. The tables were far too flimsy for heavy-weight (sic) male students and reading space was too restricted.[9] Gradually the idea of a library of distinction, convenience and comfort took shape. It would not be a mere complex of rooms housing books and journals, but a laboratory in which students would work on a topic, using many sources. Adequate working space would therefore be essential. Minimum space requirements of 3000 square feet for books, 8175 for readers, 1200 for staff and 2885 for services, a total of 15,260 square feet, were set when space schedules were drawn up in 1963.

But Llewellyn's demand that the complete Engineering School should not exceed 200,000 square feet meant that some pruning of library space was essential. Reader space and traffic aisles were reduced, a workroom for staff disappeared and the library space schedule fell to 13,045 square feet. Collins expressed his anxiety both about the present and the future:

> Even assuming the correctness of the reported U.G.C. ruling that the School of Engineering will never go above 700 students, its library will grow. Therefore I feel bound to record my alarm that no provision has been made in this already modest building for some practicable expansion scheme for the future . . . the hard fact is that a library built now without adequate facilities for development will result in complaints from Engineering staff and students later. This has already happened with the present quarters because, despite protests, the areas recommended by the Librarian in 1951 were reduced and no effective provision was made for extension. Since then, as usually happens, the demands for more and more varied library service show that the Library's full proposals at that time were not too bold but too timid.[10]

Collins returned to the idea of a basement "if technically feasible", proposed that foundations be laid to cater for an additional floor subsequently and then suggested another possibility:

> Adjust the floor levels to allow a mezzanine to be inserted (preferably above the upper floor), reduce the total ground area occupied initially, and reserve absolutely the remainder of what is now allotted for a horizontal expansion later.[11]

Those ideas were adopted at sketch plan stage and the mezzanine proved to be something of a saviour. It was aesthetically pleasing, it was a practical solution and it provided a useful demonstration of the value of a mezzanine for the planning of the Main Library. The building was actually designed in five modules, only four of which were built initially.

When it was completed in 1967, the library set a new standard for New Zealand libraries. Not only did it have 20,000 books on its shelves with space for 10,000 more; there were also 200 individual desks, some small study rooms and a seminar room as well. It was a fine example of architectural grace, functional design, tasteful furnishing and interior decoration designed to give an atmosphere of quiet dignity conducive to study. It belied its specialist nature too. Exhibitions of painting, sculpture and pottery and displays of books on a variety of topics other than engineering enhanced it. But perhaps its most striking feature at that time was the floor covering. Initially it was decided to lay quality carpet throughout the library to test Collins's claim that in addition to its inherent qualities, carpet wore better and would cost less to clean than other coverings. No New Zealand university library at that stage had carpeted floors, but Collins had seen the future in the United States and it worked. However, only half of the money sought for furnishing the building was granted by the U.G.C. and carpet would take the lion's share of it. Accordingly, after much agonising, it was decided to lay a bronze green commercial grade carpet on the ground floor only and to put a cheaper covering, brusella, on the upper floors. At least it would enable some comparisons to be made when planning the Main Library.[12] The carpet — and the test — were successful and carpet of the same grade and colour were used subsequently in the Main Library.* It was also laid subsequently on the upper floors because the brusella wore out and could not be replaced; by 1979 it was no longer made. Collins's claim has been thoroughly vindicated. Twenty years after it was laid the ground floor carpet remained evergreen.

It was not only the new building which transformed the Engineering Library: J.E.C. Shearer** Engineering Librarian from 1962 until his death in 1979) played a major part too. He was perhaps fortunate that his appointment coincided with a freeing of resources after the freezing fifties, but his quiet and unassuming manner belied a man of action with a sturdy conviction of the value of his work. His affection for books, mastery of library systems and furtherance of the science of records combined to enable him to develop an outstanding library which was subsequently shared with the School of Forestry after it opened at Ilam

*The carpet in the Main Library certainly impressed the Chairman of the U.G.C., Sir Alan Danks, when he inspected the new Library-Arts building in 1974. The carpet had just been laid, but the bookstacks were still to be erected and the floor looked not unlike a spacious, well-manicured park. Danks mounted the steps to the Library, opened the doors, whistled slowly and turned to his companion. "Caddie," he said, "hand me my driver."

**James Edward Courtney Shearer 1915-79; educ St. Albans and West Christchurch Dist HS; clerical work, Fire Service, Country Library Service 1948; appointed Engineering Librarian 1962, Associate Librarian 1973.

in 1970. Shearer was not only a librarian, but an integral part of the School and a valued contributor to many of its activities, such as the publication of the prospectus or the organisation of conversaziones and conferences. And he also made full use of his practical interests and abilities. The Engineering Library had the first photocopier, the first microfiche information retrieval system, the first microfiche copying service and the first electronic book detection system within the University. They were not mere gimmicks with which to demonstrate a librarian's awareness of technological change, but genuine aids to more efficient learning and better research. The contrast with 1960 is remarkable. Then Collins, greatly daring, decided that the old typewriter used by Miss Hale should stay on the city site and a new machine be purchased from the fund for equipping the new library.[13]

The contrast in book stock is also remarkable. In 1951, the library contained 3000 volumes, 160 current periodicals and a large number of pamphlets and other ephemera. There were also sets of British, Australian and New Zealand standards, a file of the *Engineering Index* and most of the standard abstract and reviewing journals. The collection was fully catalogued and classified. By 1960 the stock had risen to 13,000 volumes, by 1970 to 40,014, by 1980 to 68,014 and by 1985 to 87,248 volumes. In 1960 the Engineering Library had 10 per cent of the £15,000 allocated to the College Library; by 1970 it had $23,490, or 15 per cent of the total Library allocation, and spent $12,824 on books and $10,666 on periodicals; by 1980, it had 15.3 per cent of the total Library allocation of $700,000 and it spent $29,752 on books and $77,648 on periodicals; and in 1985 it had $234,805, or 14.8 per cent of the total allocation of $1,581,385 and spent $55,600 on books and $179,205 on periodicals.

Growth of that order necessitated additional space, and Shearer's successor, Heather McCarrigan, had good reason to bless Collins for his foresight. The land to the east of the new building had indeed been reserved and there was clearly an urgent need for extension into that space to accommodate the rise in student numbers, the spate of engineering books and periodicals and the need to house the growing library of the Forestry School. The faculty gave the extension high priority, but it was the needs of that omnivorous consumer of resources, computing, that finally ensured an early start. The extra "module", opened early in 1985, had a cluster of computer terminals on the ground floor, but on the upper floors 54 extra reader places and space for another 30,000 books were provided.

CHAPTER THREE

College to University, National School to Faculty

In 1958 three Bills passed through Parliament changing the names of Canterbury, Auckland and Victoria University Colleges to universities. "Whim-Wham"* commented:

> The Revival of Learning was never like this -
> Universities, three in a Day!
> As thick as Espresso Bars grow in the Street,
> New Oxfords and Heidelbergs sprout round our Feet,
> Who says we have no intellectual Elite?
> Kapai! Gaudeamus! Hoop 'ray![1]

The change may have been more shadow than substance, but it did prepare the way. When the Act establishing Canterbury as an autonomous degree-granting institution was passed three years later, the University took the major changes completely in its stride. For the Engineering School though there were more fundamental and more painful changes which in the end altered — and altered for the better — the relationship between the School and the University. They were not made without conflict; but the resolution of that conflict also brought about a coexistence that Scott and many of his successors could not have believed possible. Indeed there is no doubt that the School has fared better as an integral part of the University than when it stood alone and often aloof.

Paradoxical though it may seem, the isolation of the School at Ilam from the beginning of 1960 was one of the causes of the closer integration. The transfer to the raw, unfinished buildings and a site still littered with the detritus of building operations, cut the physical ties with a

*The nom de plume of the poet Allen Curnow, on whom the University conferred an honorary Litt.D. in 1975.

vengeance. If there was promise in the space now provided for lectures and laboratory work, gone were such other comforts as the crowded city site afforded — the quaint little library, the lunch-time concerts in the University Hall, the gymnasium and the debate over tea in the Senior Common Room; gone was the opportunity to feel a part of the city's life, to stroll at lunch-time around shops, gardens or park; and gone almost overnight were the familiar old laboratories, lecture rooms and offices that had been the School's home for more than 70 years. Even if there was a sense of relief at the prospect unfolding at Ilam, there was a strong feeling of nostalgia about leaving the city site. Propinquity had not made the School an integral part of the College; separation made the heart — if indeed there was one — grow fonder.

Stronger influences than that combined at the same time to push School and University closer. The development of the second school of engineering at Auckland may have been slow, but it meant that Canterbury could not continue to describe itself as the "National" (sometimes even "*The* National") School of Engineering, though it did not lightly drop the title. If MacElwee had been allowed his way, there would have been no change at all. Asked by Donald Bain, chairman of the Council in 1955, about changing the School's title, he said the matter had been discussed by the engineering professors from time to time and it should go before the Council for an authoritative decision. But he added:

> My personal opinion is that it is unfortunate to change the name of a group or organisation which has been in existence long enough to earn some reputation in the world under the title which it at present enjoys. The fact that the School is no longer "National" insofar as it is not the only School of Engineering in the country is of no great significance to me personally. It is a name with a reputation and it is my opinion that the matter should be weighed very carefully before the name is changed.[2]

There was no formal debate, or at least no record of one, about changing the name and diehards in the School continued as they had begun. The School ordered its own stationery and successive Vice-Chancellors noted that correspondence from Ilam was headed as coming from the "National" School. Llewellyn, who was soon to become the first chairman of the new U.G.C., was unruffled, though, as will be seen, he was to clip the School's wings in another way; but his successors occasionally expressed exasperation either at the School's stubborn refusal to face the inevitable or its seemingly endless stock of old letterheads. It all came to an end in 1968 when a new University letterhead was designed for all departments. The School and its departments were seen formally to be in the fold.

The annual prospectus was another distinguishing mark of the School. As it lost its "national" tag, and the prescriptions for its courses appeared in the Calendar, questions were asked about the need for a separate prospectus, which contained much the same information.[3] The prospectus continued though and the School, by some historical sleight of hand that baffled seasoned administrators, even managed to persuade the administration to pay for it until an economist became Vice-Chancellor. The prospectus remains, but the cost is now met by the School.

Another nagging issue remained to be resolved. A hint of future conflict about the administration of the School was given by Bain as early as 1951 when he acknowledged that provision for a School office as well as staff and student amenities would be needed in the new School. But he hoped, he said, they would not remain when all departments moved to Ilam as that might lead to segregation of the School in perpetuity.[4] As has been seen, the student cafeteria remained; so too did the curiously-titled Bursar, who was neither treasurer of the School nor the holder of any bursary. In 1949, long before the passing of the Human Rights Act, the College intended to advertise "for the position of Bursar (Male) in the School of Engineering",[5] but wiser counsel prevailed. The best candidate for the position was a woman, Ailsa M. Comerford, who had been an assistant in the office. She applied when the word in parenthesis was removed from the advertisement and was appointed.

When planning the new School, Rastrick told Hulme that it worked as a special unit, "a special School of the New Zealand University. It keeps all its own records of students and due to the nature of the course, practical requirements etc. the administrative work for students is greater than in other faculties,"[6] The Rector reported subsequently that the Council had questioned the large amount of space planned for the office and suggested that Dean and Registrar should discuss it. The implied criticism of the office generated uncertainty subsequently and the Bursar (she had become Mrs Unwin) had some questions for the new Registrar, G.G. Turbott* about her future when the Registry moved

*Turbott, Garth Graham (1919-). Educ Takapuna G.S., A.U.C. (B.A., B.Com); J.P., FCA; 2 NZEF N.Africa, Italy (MC); asst. reg. AUC 51-56; Registrar CU 1957-78.

to Ilam. No firm policy had been established and no promises were made. But it became clear as the reunification of the University at Ilam approached that closing the Bursar's office would merely inject a discordant note into the growing harmony between the School and the University administration. Some engineers had been heard to mutter that the Registrar wanted to close the office simply to increase his own empire. If that argument was difficult to sustain, there was more substance to the claim that the Bursar did much more than act as a faculty administrator. She also kept student records, found and allocated vacation employment in appropriate industries for students, checked their workshop experience, arranged the required first aid courses and even sold special stationery to them. Mrs Unwin resigned in 1963 and her successor, David Thomson, was told he was not so much a bursar as a faculty administrator, even though he would sit behind a door marked "Bursar" and sign correspondence under that title.

In fact, few economies would have been achieved by closing the office; any savings would have been outweighed by the ill-will a closure would have generated at a time when harmony was important if a sense of community was to be encouraged at Ilam. But from time to time there were reminders that the position could be an irritant. In 1975, for instance, the Dean (Professor H. McCallion) thought it was "incongruous" that the senior secretary in the bursar's office should have a manual typewriter when departmental and Registry secretaries used electric typewriters.[7] The Registrar agreed that the office could have an electric typewriter on the ground that the office "forms part of the Registry of the University".[8] When University committees were reorganised subsequently, the Bursar was given secretarial responsibility for two of them, a step that put him on the same footing as other faculty administrators.

The changes might have been more symbolic than real, but they inexorably pushed the School into a different relationship with the University in the early sixties. Instead of being a ward of the State, fostered by the University, it became a fully adopted and often favoured son. To be absorbed into a family may be comfortable and secure, but it may also lead to a loss of independence. It was Llewellyn who ensured that the School would be adopted *de jure*, though his motives were not exactly fatherly. As Rector and Vice-Chancellor, he had battled long and hard for completion of the School , its equipping and furnishing on an unprecedented scale and for almost immediate extensions while simultaneously planning the much large science buildings. His departure was, as the University History puts it, a double loss to Canterbury. Not only was he an outstanding Vice-Chancellor, but also in his new position he quickly made sure the departmental back door he had so assiduously used in Wellington was locked against other Vice-Chancellors, including his successor.[9] And it was double-locked against special schools, which in the past had trodden well-worn paths to Ministers of Health, Agriculture or Works. The chairman of the new U.G.C., established by Parliament on the recommendation of the Hughes Parry Committee, decreed that the special schools of medicine and engineering would no longer receive special grants quite separate from the Arts, Science and General (A.S. and G.) block grants made to each university. From the end of the 1960-64 quinquennium the resources which sustained the Engineering School would have to be obtained from the University in the face of competing demands from five other faculties.

Llewellyn had given general warning of his intentions at informal discussions, but the first document in which it was formally spelt out appears to be the letter from the U.G.C. secretary, R.G. Rowley, giving details of the 1965-69 quinquennial settlement. "You will see," he wrote "that from 1 April 1965 there will be one grant for the whole of the University instead of separate grants for A.S. and G. and for the special schools."[10]

Llewellyn's reason for that step was to ensure that in future every sector of the university system would negotiate with the U.G.C. and not with any Minister or department. It was necessary if the U.G.C. was to be anything more than a pale copy of the Senate Grants Committee, but the ukase had an almost dramatic effect at Canterbury. The Engineering School was finally demoted from the status of special school to mere faculty. Even had it wished to do so, the School could no longer stand aside from the mainstream of University government, administration and activities.

Even though it had been in existence since Scott's retirement, in 1924, the faculty did not have quite the same prestige once it was shorn of national school status, a point underlined in a curious brush in 1962. Having settled into Ilam, the faculty was anxious to use its new buildings and equipment to encourage graduate study and it was critical of some major Government departments employing engineers. Indeed it resolved to ask the Liaison Officer with secondary schools to draw the attention of school careers advisers to restrictions placed on Government engineering cadets who wished to undertake postgraduate work.[11] C.S.

Planck, the Post Office engineer-in-chief, and a member of the faculty, was clearly annoyed.

How far (he asked the Vice-Chancellor) could the faculty direct the Liaison Officer?[12] Pownall replied that faculty decisions had to go through the Professorial Board and usually the University Council and the University was a long way from carrying into effect those decisions.[13] And the Vice-Chancellor reminded the Dean that the faculty had no legal entity. "It is essentially a piece of internal administrative machinery," he said . . . "it would seem to be extremely limited in its powers to send letters outside the University."[14]

The faculty took the point. It was not caught offside again. The School did not, indeed could not, object to being mustered into the University's financial fold. What it could do — and indeed did — was to ensure that its claims for a goodly share of the increased block grants of the sixties were heard. Engineers were well represented on every important University committee; they spoke with one voice on the Professorial Board; and if practising engineers largely disappeared from the University Council, one and sometimes two academic engineers could usually be found there. Certainly there was no reduction in the proportion of the block grant that went to the School. Engineers no doubt joined on occasion in expressing dissatisfaction at the general level of funding for the whole academic enterprise, but there is no evidence that the School was unhappy with the proportion of those funds which flowed to it compared with other departments. Loss of the direct grant meant a loss of independence rather than a loss of money. It also meant assessment of engineering requirements and aspirations by academic colleagues also seeking a share of the block grant cake. If there was reluctant recognition of the fact that an engineering student cost much more to educate than any other Canterbury student, though not nearly as much as a medical or veterinary student, there was no intention of allowing the School to drain resources to the detriment of other departments. Resolving such conflicts created tensions — but should not tension be a permanent condition of the University? — resulting finally in a *modus vivendi* that drew School and University into close partnership.

The development of that harmony is perhaps best illustrated by debate within the University on proposals from the Engineering faculty for inclusion in the University's quinquennial submissions to the University Grants Committee. Initially, of course, those submissions were made to the Senate's Grants Committee, established in 1948 to enquire into the financial needs of University education and to advise the Government about them. Oddly enough, that committee, which had no statutory powers, gained a vital advantage for the university colleges — or rather for the A.S and G. components of them — by persuading the Government to accept quinquennial funding, even if the principle was hedged about with a number of conditions. A formula based on predictions of student numbers determined the level of the grant; and because numbers rose, so too did the level of the grant. But the Engineering School, along with other special schools, was excluded from this funding. Direct grants for the schools were worked out according to needs, not always the needs of the schools but also national requirements for qualified doctors, dentists or engineers. And though enrolments influenced the scale of the grants, money lagged behind roll increases. Moreover, the committee had little of the expertise its successor would call on and some of its decisions about claims for new and specialised equipment seem now to be perverse.

Quinquennial funding began in 1950 but it is difficult to determine the precise proportions of resources that flowed to engineering compared with the College. From its grant the School paid an agreed sum towards administration as well as other "housekeeping" costs — cleaning, heating and lighting, for instance. In 1960, when its student roll was 14.3 per cent of the total roll, the School had 24.5 per cent of total University income. In the last year (1964) of its separate funding, the School had 14.1 per cent of the University roll and 20.4 per cent of total income. Direct departmental costs for each engineering student were calculated as being the equivalent of $735 in that year. The average cost of students other than engineering students was the equivalent of $357. Two decades later, in 1985, analyses of costs which took into account all of the University's recurrent expenditure showed Engineering students as being by far the most costly students in the University. Costs per equivalent full-time student (excluding expenditure on bursaries and allowances and the capital costs of buildings and equipment) were: Law $2,470; Arts $4,818; Science $7,621; and Engineering $12,078. The average total expenditure per equivalent full-time student, throughout the University was $5,927. The figure for each Engineering student was more than double the University average, more than half as much again as the figure for Science students, more than twice that of Arts students and five times that of Law and Commerce students.[15] Clearly the School did not suffer from formal integration.

All other measures confirm that finding. In the fifties, the ideal staffing arrangement, sometimes likened by the irreverent to a rugby scrum, was considered to be three professors, five senior lecturers, five lecturers and three assistant lecturers for a team of 16 (not 15) staff. Because the School could not — or said it could not — recruit academically qualified engineering staff at assistant lecturer level, it sought a proportion of two professors, one associate professor, seven senior lecturers and six lecturers. The 3:7:6 combination looked formidable enough; but what mattered more was the quality of that staff and the ratio to students. On both counts the School fared well.

The School sought substantial increases in its grant for the 1960-64 quinquennium. The new buildings required more cleaning, heating and lighting; its staff:student ratio had worsened from the 1:9.1 in 1954, to 1:12.7 and it sought an improvement to the ratio recommended by the Grants Committee of 1:9.5. It also wanted the existing ratio of one technician to 2.9 academic staff to be reduced to 1:2. In 1960, the School had a full-time permanent academic staff of 36, of whom four were professors and one an associate professor.

Ten years later the academic staff had doubled in size and the roll had risen to 834. The number of professors had doubled and there were seven staff in a new ranking of reader. Only two of the 36 staff in 1960 held doctorates; whereas in 1970, 18 of the 72 staff were qualified at doctoral level. Staff numbers rose to a peak of 82 in 1975, with 12 professors and eight readers. By 1985, as "disestablishmentarianism" became known at Canterbury as something other than the longest word in the language, the number fell to 75, including 14 professors and eight readers. Fifty-seven of them held doctorates. In 1985, the overall staff:student ratio for the School was 1:9.1 (compared with 1:14.3 throughout the University). The Science Faculty undertook some 15 per cent of all the teaching of professional engineering classes.

Similarly the School seemed — at least to other hard-pressed departments — to bulge with technicians, a legacy, at least in part of MacElwee's untiring work towards the establishment of the New Zealand Technicians' Certification Authority. When he first arrived in New Zealand, MacElwee saw the need for a trained group between professional engineer and artisan capable of freeing the former from the performance of routine engineering tasks and testing procedures.[16] By 1960 his advocacy, and that of his colleagues, for technical assistance had resulted in the employment of 11 technicians, with one appointment pending, a ratio of one technician to 3.27 teachers. Ten years later the number of technicians had risen to 49, with three positions unfilled, and in 1985 the total was 65, close to the 1:1 ratio regarded as desirable by all interested parties but the U.G.C. which required a reduction of 50 non-academic staff a year for three years starting in 1981.[15] Universities such as Canterbury, which had put more resources into academic support staff than others, were forced to accept correspondingly greater reductions. But the axe did not fall on technical positions in the School and in 1986 the number was the same as six years before. Typically Civil Engineering had 33% of technical staff, Mechanical 27%, Electrical and Electronic 25% and Chemical and Process Engineering 15%. Technicians were increasingly better qualified too.

Another measure of the School's standing within the University was its success in gaining acceptance of new developments either through proposals for inclusion in the University's quinquennial submissions or by winning Professorial Board approval for new courses that might be fitted within the existing framework. The School's academic politicians, and especially Hopkins, could scarcely be described as tentative in presenting faculty proposals, but they did not, because of their separate grant, need to compete with other faculties until the third quinquennium. By then the School was soundly based in new buildings, some of which were being extended, and with a considerable amount of new equipment; and it was not until the 1969-74 quinquennium that substantial new developments, including a department of engineering mathematics and development of what was described as a university computer "sub-centre" were sought. Computing, as will be seen, had its beginnings in the School and the engineers quickly appreciated how the IBM 1620 and its successors provided the means for constructing mathematical models and conducting experiments on a scale that had hitherto not been possible. More, computers would enable engineers to simulate the performance in advance of a complex engineering enterprise or industrial process for a variety of design data far beyond any possible physical model. Staff had observed overseas a new emphasis on mathematics, numerical analysis and programming, on the solution of engineering problems on digital and analogue computers and on the application of computers to process control, engineering design and the simulation of complex systems. To produce engineers who were better mathematicians, the faculty claimed, the School needed mathematicians who could become better engineers.[18]

The faculty also sought a new course in chemical process technology and the introduction of building technology into civil engineering as well as the establishment of a special master's degree to expand engineering knowledge and skills by advanced study rather than by research. But initially none of the proposals got very far. A special Professorial Board committee on the University-wide submissions ruled that the proposal for engineering mathematics would be regarded merely as an alternative way of organising mathematics in the University and thus a legitimate charge on the grant rather than as a fundamental new development. Chemical process technology and building construction were also part of normal growth; and as for the proposed postgraduate course in civil engineering, it was not only a matter of normal growth but one that was also of concern to many other departments.[19] In fact Electrical Engineering, with very little fanfare, was already offering an M.E. by examination in five fields of study. All departments now offer a similar degree in many different fields — all without recourse to the U.G.C. Little provision was made by the U.G.C. for any major new academic developments in that quinquennium, but it is not without significance that the faculty's requests were met, at least in part, by the University in subsequent years. The Professorial Board might not accept the division of mathematics, but it would provide for diverse needs within that framework.

Professor Frank Henderson, a graduate of the School, who had taught civil engineering since 1952, first as a senior lecturer and since 1964 as a professor, was an ardent supporter of the proposal for an engineering mathematics chair and his departure for a chair of civil engineering at the University of Newcastle, N.S.W., in 1968 was seen by some observers as an expression of his annoyance at the outcome. But he made it clear that his reason for leaving New Zealand was his dissatisfaction with the "unfair" taxation of the professional middle class. In a final lecture to accountancy students he said he had come to the conclusion that New Zealand was not worth the effort "and my advice to you is that you will follow me out of this country as soon as you can."[20] The *Press*, in an editorial headed "The Pied Piper of Ilam," admonished him for giving "poor advice" to such an audience.[21] Professor B.A. Woods* who had specialised in engineering mathematics at the

*Woods, Brian Anderson, 1931 –; educ Waitaki B.H.S., CU, Coll of Aeronautics, Cranfield; B.E (Mech), DC Ae.,FIMA; commissioned RNZAF 54-59, sen.sci off & prin.sci off, R.A.E. Farnborough 59-65; lect. in fluid mech & app.maths Un. of Leeds 65-69; prof of maths CU 1970–.

Royal Aircraft Establishment, Farnborough, was appointed to a third chair in mathematics in 1970. Provision for computing, the growing cost of which disturbed administrators as much as its promise delighted practitioners, was not so readily resolved, but the current solution is not very dissimilar from what was sought — a function more of the development of micro-computers than of academic foresight. By the time submissions were in train for the 1985-89 quinquennium, the School's priority was the establishment of a computer-aided design system. The computing world has not won acclaim for the simplicity of its terminology and some academics might have wondered quite what was meant by one proposal — "a central super-mini-computer" with "semi-intelligent work stations linked to it by a high speed network (ethernet etc.)"[22] Those who were puzzled by the description could take little comfort in a belief that computer hardware and software were continually becoming cheaper. The cost of establishing a 40-station CAD system was put at $1,182,600, together with annual maintenance charges of $120,000, staff costing at least $55,000 a year and an additional 300 square metres of space in the School. But there was no doubt a computer-aided design system would be established. CAD has rapidly become an essential tool of the engineering profession and students must be as steeped in it as in the second law of thermodynamics. The University and U.G.C. recognised it as an inevitable if costly development and it awaits only the money.

The School, in particular Chemical and Process Engineering, was also closely involved in the major submission for the quinquennium, the development of biotechnology within the University. Biotechnology, described as investigation of the integrated application of microbiology, biochemistry and process technologies to biological systems, is considered to hold as much commercial promise as the revolution in micro-electronics with applications in food production, alternative energy sources, substitution of petrochemical feedstocks, waste recycling, pollution control, medicine and pharmaceuticals.[23]

All these factors indicate that while individual departments prosecuted their own cases purposefully, the University as a body made generous allocations for Engineering . Nowhere is this shown more strikingly than in the provision made for teaching and research equipment. Individual engineers and departments, industry and the profession contributed and the setting up grants made when the new buildings and their extensions were occupied also helped; but the fact remained that the School possessed more than half the teaching and research equipment in the University. In 1978 the replacement value of all the University's equipment was estimated at $14 million, with Science and Forestry accounting for $6,500,000, Engineering for $7,300,000 and all other faculties for $300,000.[24] Subsequent inflation might well have doubled the figures, but they were scarcely a source of satisfaction as much of that equipment was obsolescent and its replacement had become of prime concern to both School and University.

CHAPTER FOUR

Elusive Expansion

An Air New Zealand direct flight from Dunedin to Auckland passes silently and, at 30,000ft, almost invisibly over the University; when conditions are right the passage of the aircraft is revealed by a startlingly white vapour trail. To Canterbury's engineers, the passage of the Department of Mineral Technology at the University of Otago to the School of Engineering at the University of Auckland in 1985 seemed similar. Its flight path had been surprisingly set for Auckland, bypassing the site proposed for it at Ilam and leaving no tell-tale trail. If there were sounds of anguish in Dunedin that a department established as long ago as 1879 could be snatched from its fold, they were accompanied by discreet expressions of exasperation in the School of Engineering, which had been at one stage front runner in the choice of a new home for the migrating department.

It was the second occasion since the move to Ilam that departments which would have been profitably associated with the School had gone elsewhere. But as past argument over the location of engineering and forestry showed, the New Zealand university system is not seen in its best light in the competition for special schools or new departments. Parochial considerations and political pressures tend to outweigh academic profundity, despite the frequently expressed wish of the universities to co-operate to achieve the best location for teaching and research in the subjects concerned. But if Canterbury had a good case for adopting mineral technology, disinterested observers were even more surprised that the second School of Architecture was not established alongside the Engineering School.

The New Zealand Institute of Architects raised the possibility of a second school of architecture with the Senate in 1957 when it proposed that there should be 300 architects per million of population, a figure that the existing school of architecture at Auckland could not hope to meet. The institute proposed that any second school should be allocated to Victoria University of Wellington, a claim that rested more on past default than on common sense. Victoria, a late starter in higher education, had looked in vain for a prestigious special school. They had all been appropriated by Otago, Canterbury and Auckland and 60 years later Victoria felt, if not naked, then improperly dressed without one. Architecture seemed, at last, to offer a prospect of professional education; and Victoria as well as Wellington architects, found rather less than compelling arguments for siting the school there. The capital had the largest architectural offices in the country and the ratio of architects to population was highest there, so the pool of architectural and structural engineering skills would far exceed that in Christchurch. Wellington had a higher population and there was more building. Conspicuously absent were good academic reasons for choosing Victoria.

The Senate asked the colleges for their views. All but Victoria agreed that when a second school was established it should be at Canterbury because of the desirable association with the existing Schools of Engineering and Fine Arts. The Hughes Parry Committee came to a similar conclusion. Local architects even applied to the Senate's Grants Committee for authority for courses in architectural subjects to be established[1] and the *Press*, commenting on the "persuasive" arguments, said Christchurch had an excellent claim to the new school.[2] Special schools might have been allocated for wrong reasons in the past; surely one would not be allocated now to Victoria because it had missed the bus.

The new U.G.C. was not prepared immediately to consider the establishment of a second school and suggested that courses for architectural draftsmen might be looked at.[3] The Quantity Surveyors' Institute also sought to join the party; and Frank Lu at the same time proposed the gradual development of a degree course in building science, based on civil engineering, a proposal warmly supported by Hopkins. He foresaw the possible development of building technology, structural engineering and architecture and hoped this sort of "cross-fertilisation" might ultimately produce structural engineers well versed in the other disciplines.[4] From all this Pownall prepared a detailed submission, suggesting a school providing two options. The first would produce a traditional architect, though it was proposed graduates would have had considerable exposure to arts, science, commerce, law and forestry experts as well as to engineering and fine arts. The second option would be a graduate oriented towards the civil engineering-mathematical aspects of architecture. A second school producing a new breed of architectural engineer (or engineering architect) was, he said, a matter of national importance because New Zealand was suffering from not having this type of architect.[5]

The submission did not fall on stony ground exactly, but neither did it take root. The U.G.C. agreed to support in principle the establishment of two further faculties of architecture, including ultimately (fateful word) one at Canterbury. The other (and, as it turned out, more immediate) faculty would be at Victoria. Canterbury was invited to begin planning a school on the Ilam campus for an output of 40 graduates a year and it was suggested that construction might be "phased in" when the major library and arts buildings were well advanced.[6] There were only a few crumbs of comfort in that decision and the Professorial Board and Council mulled it over before advising the U.G.C. that the next Faculty of Architecture in New Zealand "would best be located at this University."[7] But try as it might, Canterbury could not obtain a U.G.C. definition of "ultimately".

Time passed; a new Vice-Chancellor (Professor N.C. Phillips)* was preparing Canterbury's quinquennial submission in 1968. Should Canterbury (he asked a new U.G.C. chairman, Mr A.J., later Sir Alan, Danks) argue the case again or was the past approval in principle sufficient? Danks advised him that while Victoria would get the second school, the matter had been deferred pending construction of better accommodation for the existing Auckland school. It was obvious to Phillips that a long time would elapse before a Canterbury school became possible and there was no point in working on another submission.[8]

But local architects, convinced that Canterbury was an ideal location for architecture because of the inter-relationship of disciplines, were unwilling to pigeonhole the proposal. The school might not have a high priority, but why could a course in architecture not be established in the Engineering School in conjunction with building construction?[9] Such a course might be a sort of architectural intermediate (though would Auckland concur?) and it would certainly be the thin end of a wedge

---

*Phillips, Neville Crompton, C.M.G.; b 1916; educ Dannevirke HS, Palm N. BHS, CUC (M.A.hons); hon. D.Litt; jnlist *Sun & Press* 19 32-38; pg. schol Merton Coll, Oxford 1938-39; gnr to major RA Sec W.W. N.Africa, Italy; history lect CUC 1946, sen lect. 1948, prof. 1949-66; V.C. 1966-77.

for Canterbury. Discussions were held. The Institute of Architects, while endorsing consolidation of the Auckland school as its first priority, was prepared because of stalemate over the second school at Victoria, to investigate other means of alleviating the shortage of architects. There was a suspicion that the U.G.C. might prefer architectural training to be relegated to technical institutes. But Phillips pointed out that the Engineering School already had restricted entry and the University would need a capital grant to extend accommodation and employ lecturers for such courses. So while the school remained in the background of successive Canterbury quinquennial submissions, there was little point in trying to move heaven or earth or both in pursuit of architecture, especially after a second school became a reality at Victoria in 1975.

Negotiations over mineral technology began in an entirely different way. Though the two engineering schools treasured their independence, they were not unaware of the benefits of collaboration. Discussions resulted in an Auckland proposal for a review of engineering education and terms of reference were discussed in 1979. Both schools subsequently asked the U.G.C. to assess whether the number of engineering graduates was meeting national needs, whether the resources of the schools were at an appropriate level and whether degree standards were meeting the needs of the profession. Professor A.G. Williamson, who was Dean, reported to the faculty in 1980 that the Vice-Chancellor, Professor A.D. Brownlie,* would head a committee, the other members of which were Messrs J.C. Fair and J.H. Ingram, to review engineering education for the U.G.C.[10] The Brownlie Committee dealt with six other aspects of higher education, but its investigation of engineering was especially swift and incisive and its four major recommendations were rapidly heeded. Examine the engineering degree structure, especially the purpose of the intermediate year, the committee advised; provide more university resources to meet an increased demand for engineers; establish a committee under the U.G.C. to guide engineering and technological education; and, finally, strengthen the Mineral Technology Department at Otago, particularly by associating it with an engineering faculty.[11] The department had grown from Otago's old School of Mines, which lost its prime purpose when the exploitation of most New Zealand minerals fell off. Indeed I.D. Dick, the Under-Secretary of Mines,

*Brownlie, Albert Dempster; b Fiji 1932; educ N.Plymouth BHS, AUC (M.Com); Treasury 1954-55; economics lect AU 1956-64, prof. CU 1965-77, asst V.C. 1974-77, V.C. 1977-; chmn Monetary and Econ Cl 1972-78.

recommended in 1968 that the department be transferred to one of the engineering schools, preferably Canterbury; and in 1976 an Otago investigation concluded that it was unlikely any other university would want the department.[12]

But both Auckland and Canterbury were now very interested. Submissions emphasising the undoubted advantages of each engineering school flowed into the U.G.C. sub-committee on engineering that Brownlie recommended. A third submission from the defendant, Otago, sought retention of the status quo, with some improvement. There was talk of "offshore" association on the lines of the Christchurch Clinical School. The limitations of the Auckland site, it was said, told against it. There was space enough at Ilam — though not in the existing School — and a new building of 600 square metres with shared laboratory extensions of 300 square metres was proposed. With equipment, offices and books the cost was estimated at $2,360.000.[13] But the U.G.C. contested the Canterbury proposal for 20 graduates a year and the estimates were reduced to cater for half that number. Even so, at that stage in the inquiry, Canterbury was favoured by a small majority of the sub-committee.[14] Yet the climate appeared to change after the full U.G.C. visited Ilam in October 1983 to discuss the University's quinquennial submissions. A new Dean, Professor D.C. Stevenson, emphasised the Faculty's principal submission, which was for finance for computer aided design in the School. The Science Faculty also supported the transfer, but its priority was the introduction of advanced biotechnology. An impression, difficult to dispel, that Canterbury was less than enthusiastic about mineral technology, spread abroad, even though the Engineering faculty strongly supported the transfer to Canterbury if it was invited to take it in.[15]

An invitation was unlikely; more could be expected from a sound case and perhaps the application of political pressure. Canterbury's case lacked nothing in logic or confidence but it was not — perhaps too high-mindedly — backed with political muscle. And there was a slightly uncomfortable awareness that the near extinction of operations in the Westland coalmining industry vitiated the argument. The ready access from Ilam to all types of exotic and native forests had been persuasive in the case for establishing the School of Forestry at Canterbury in 1970. Access to developing coalfields favoured Auckland's claim.

At the end of 1983 the U.G.C. expressed its intention to associate mineral technology with a school of engineering. By March 1984 it had decided that Auckland would be the school. The grapevine hummed. Otago fought a determined rearguard action in the belief that though the U.G.C. would provide the advice, the final decision would be made by the Government. Alas, rumour was substantiated by formal advice in October that the U.G.C. would cease funding the Otago course from the start of 1986 and establish mineral process engineering and mineral technology at Auckland. Political pressure resulted in a stay of execution and an independent review. Canterbury expressed its enthusiasm again and offered a solution to accommodation needs: the speedy construction of a proposed heavy engineering laboratory of 1000 square metres at a cost of $1 million and subsequent provision for permanent headquarters for the department.[16] It was all to no avail. The flight north bypassed Ilam. Whatever forces attracted it to Auckland could not be discerned; there was not even a vapour trail to mark its passage.

The remnants of inter-provincial rivalry could also be discerned in attempts to resolve the other recommendations of the Brownlie committee. An ad hoc U.G.C. sub-committee set up to examine the recommendations in detail threw engineering planning into some disarray by drawing a parallel between the number of engineers within a country and its economic development and suggesting that New Zealand needed more engineers to gain, as rapidly as possible, some sort of economic parity with developed countries. It was certainly a more sweeping approach than the questionnaires circulated from 1969 to 1979 by Stevenson and J.B.C. Taylor, a senior lecturer in mechanical engineering, to assess the demand for engineers; and it was scarcely surprising that it should meet some stiff opposition. The working party of the sub-committee said that if the proportion of engineers to population was to rise significantly, the Auckland School of Engineering should be expanded for 320 additional students immediately and a similar expansion, to be ready for the 1989 intake, should be undertaken at Canterbury. More planning for a third school at Massey University should begin so the first intake could enrol in 1992. In the long run it was proposed that the third school could have 1200 students too.[17]

Professor W.L. Jones, then Dean of Science, pointed out practical difficulties in accepting the correlation. The United States, with 70 times New Zealand's population, produced about 70,000 engineering graduates a year. Scaled appropriately to population, that would result in an output of 1000 New Zealand engineering graduates a year. In 1984 both Schools produced a total of 532 graduates. From where would the additional students come? And in any case a correlation between economic growth and numbers of engineers could also be matched by similar correlations for the other learned professions and no doubt for history or political science for that matter.[18] Dr J.A. George, then Dean of Commerce, thought the forecasts should be regarded with considerable suspicion.[19]

The Engineering Faculty concluded that it would be "inappropriate" to extend the Auckland school first , especially departments which were not then filled, and recommended expansion of both schools "immediately and simultaneously" though taking into account the pressures on individual departments.[20] An ominous rumble was heard from Wellington. When a third engineering school was required, said the Victoria University Council, it should be established there.[21] And the Canterbury University Council, diplomatically pragmatic, disowned the wilder engineering enthusiasts. Believing that the forecasts about both the future demand for engineers and the supply of engineering students were subject to a wide margin of error, it recommended that the engineering schools should be expanded — simultaneously — only after there had been a sustained rise in enrolments.[22] Indeed the U.G.C. was told quite clearly that the construction of a Law-Commerce building at Ilam should have a higher priority than a major engineering building — "here or elsewhere."[23]

The U.G.C. working party did not, however, labour in vain. In fact, its investigation of the principal Brownlie recommendations precipitated an evaluation of the B.E. degree that led to substantial reform. Its members faced what seemed initially to be an intractable problem: the heavy workload carried by engineering students could not be increased; yet, as knowledge continued to expand, the content, especially the technological content, of the degree had to be increased; and extension of the basic degree course beyond four years would not be welcomed. But a solution was found. To some observers it bore a resemblance to a five-year course in disguise, though only, it must be acknowledged, for students with relatively modest entrance qualifications.

The starting point of reform was the intermediate year, stuffed with mathematics, physics and chemistry. Could not that content be reduced, perhaps by half, and replaced with engineering fundamentals, giving those who passed a head start in the professional years which followed? The most able students already skipped the intermediate year anyway; junior scholars and those with high bursary marks went directly into the professional course. Of the remainder, a considerable proportion were forced to repeat at least some of the mathematics and science they had studied in the seventh form.

The performances of both groups of students were analysed and assessed. There was no difficulty about the top students. Junior scholars had a 100 per cent pass rate in engineering and entrants with A bursaries had a pass rate of about 85 per cent. But, as in most disciplines, those with lower entrance qualifications fared more poorly: only about 45 per cent of B bursars were completely successful; and students with Higher School Certificate and University Entrance as their highest qualification had varying but undeniably lower pass rates.[24]

Mulling over the figures, the working party discerned a resolution of the problem. The best students were already selected for direct entry. Why not select also for the modified first year course those whose entry qualifications indicated that they would probably be successful? And because it would be "an intolerable injustice" if a student with U.E., H.S.C. or lower bursary marks were denied entry to engineering, why not establish a preliminary year-long course of mathematics and science to prepare those passing it for the modified first year? To those who objected that this represented a five-year degree, from which some candidates would be exempted an initial year, or even two years, there was a ready answer. Some 28 per cent of all students already took more than four years to complete the B.E. degree and more than 70 per cent of those with a B bursary took five years or more.[25] The reform not only recognised this unpalatable fact, but did something positive about it. Most of the failures that caused prolonged degree studies occurred in the first year. The redesigning of the first stages of the course would improve the prospects and performance of new entrants, especially those who were finding the initial hurdle too high.

The Engineering Faculty had no quarrel with those conclusions; indeed it had contributed to them. But a faculty committee established in 1983 to recommend a new course of four engineering years in place of the intermediate and three professional years found it was not all plain sailing. The first year engineering course had to allow a good student to cope and yet permit modified courses or direct entry to outstandingly good students. Students who took the preliminary science year would be able to postpone a decision about the branch of engineering they wished to study, but if they decided at the end of that year not to enter Engineering School at all, could they obtain full credit for their work towards a B.Sc? And what of students who wished to complete their first year at other universities? The questions are still being resolved in 1986 but not without hiccups.

Also unresolved were the resources required for teaching the new first year course. By 1985, student enrolments exceeded the designed capacity of the School (700) by nearly 200. The reforms might increase that number and they would certainly spread more students about the School, putting additional pressure on lecture rooms and laboratories. Ambitious plans for the construction of five-storey blocks, similar to that of Chemical and Process Engineering, for the other three departments remained firmly on drawing boards.

Nor did the reforms quite touch on another Brownlie recommendation — the introduction of more management, environmental and socio-economic content in to the B.E. course. The faculty was pleased with the prospect of reducing both formal contact hours and the excessive workload of engineering students when the new course began and it had no intention of adding anything from other disciplines to the educational baggage of B.E. graduates. The School had resolved itself into departments 60 years before and its common first professional year eventually disappeared. The specialist trend continued as pressures grew for more specialised and more advanced material in the first degree. The reforms acknowledged that the degree would benefit from sitting on a broader base; but that base would be engineering fundamentals and not topics such as management, commerce or law with a dash of architecture or town planning. The draft prescriptions for the new first year engineering course show it will contain engineering-science subjects and an early introduction to professional engineering subjects. Engineers might deplore in private the apparent capture of key management positions in industry and commerce by accountants, especially those trained in the techniques of management; but engineering students who planned to acquire those management skills would have to do so after their first degree. The U.G.C. working party believed New Zealand should greatly strengthen its use of engineers in senior management and executive positions and cited the experience of industrially developed countries to support the belief. In Japan, for example, engineering was the largest subject in higher education and in North America, Europe and Scandinavia an engineering degree was seen increasingly as a passport to management positions.[26] Perhaps the Jones analysis annulled that argument too for it was not pressed.

But engineering was not exactly narrowly based at Canterbury. Some undergraduates included the study of non-engineering subjects in their courses. Graduates did well in the D.B.A. and M.B.A. courses as well as in the M.Sc. in geology or resource management. And, as will be seen, engineering research students worked in a wide range of disciplines, including psychology, medicine, forestry, chemistry, music and mathematics. The B.E. clearly enabled students to spread their intellectual wings wherever their academic interests took them.

CHAPTER FIVE

Leading the School

If Canterbury's engineers demonstrated a penchant for innovation and ingenuity in their research, they tended to be more conservative in the administration of the School. The Brownlie delegation of decision-making to departments and the requirement for consultation within departments on appointments, promotion and (among other matters) the selection of their leaders, were major University reforms. The appointment of readers and senior lecturers as heads of departments even resulted in an inquiry by the Professorial Board into the role of the professoriate. Did professors, in fact, profess any more?

There were few such doubts in the School of Engineering. Professors remained at the helm and there was no suggestion that heads — of departments — should roll. Those who reached the dizzy height of a chair wanted to put something back into a system that had served them so well and were prepared to accept a term as head if the head wanted a change. In any case, readers and senior lecturers in the School were usually so involved in research that administering a department held no attraction for them at all. And for 60 years after Scott's retirement the Dean of the School was also a professor. Only in the School's hundredth year did that change.

From 1924 to 1964, the three-year term of the Dean of Engineering circulated sedately among the departmental heads, whose status was not much less than that of the god professors of old. Only Cull, as has been seen, declined office. As the roll expanded, the job grew in complexity and the prospect of appointment lost such attractions as it once possessed. Persuading a head of department — always a busy professor — to accept the position became increasingly difficult. In 1971 a faculty working party investigated alternatives, including the appointment of a professor who was not head of a department. Chairs had proliferated since Henderson's appointment to a second chair in 1964. Appointment of an associate dean and delegation of some of the Dean's duties to senior academic staff and the bursar were also considered. But the faculty preferred another proposal: establishment of a full-time Dean of Engineering.[1] It was a rather surprising proposal, for a full-time Dean might be expected — if experience elsewhere was any guide — to tread on departmental preserves. None of the other faculties supported the move and it stalled. But the engineers did break with tradition in one major respect by agreeing that professors who did not head departments could be appointed Deans and that an associate would be appointed from the same department.[2] The associate was envisaged as an assistant dealing largely with the more tedious aspects of the Dean's work and not as a deputy — the previous Dean took that post if required.

The hesitation of the other faculties about supporting a full-time Dean is understandable, but it is true that the Dean of Engineering has rather more to do than his decanal colleagues. In addition to dealing with the minutiae of academic administration, chairing meetings, representing the faculty on the Academic Administration Committee, reporting to the Professorial Board and co-ordinating quinquennial submissions, the Dean of Engineering is also required to initiate and lead the planning of the academic and physical development of the School, to maintain close links with industry, to help ensure places for student practical work and to work closely with the Institution of Professional Engineers of New Zealand at both local and national levels. He must do some entertaining, encourage and support Ensoc and be the principal liaison among the engineering departments. And he also chairs meetings of the heads of the departments, an unofficial but very effective management committee for the School. Small wonder that heads of departments were reluctant to become Deans as well, especially when the 1971 working party recommended a five-year term. Twice (in 1977 and 1983) it proved very difficult to persuade a professor to take on the task. Those heavily committed to research simply declined; a Dean is seldom able to work in a laboratory.

Fourteen years later, in the wake of Auckland's coup with Otago's Mineral Technology Department, the faculty again investigated the merits of a full-time Dean as a means of giving greater continuity of purpose and policy. A sub-committee instead recommended that more resources — it meant money — should be provided to allow the Dean more time to be Dean. He should be freed from most departmental teaching and administration; his department should be recompensed; and he should have a budget of his own and an office of high standard outside his department.[3] One significant change that was adopted brought the School into line with the other faculties. The Dean no longer needed to be a professor, but simply a senior member of the faculty.[4] If that was seen as a stopgap solution, the first non-professorial Dean, Dr A.J. Sutherland* made it clear that he would consider a five-year term if the faculty wished to change the rules. But an attempt to extend the Dean's term stalled subsequently at the Professorial Board.[5]

A full-time Dean might be the logical conclusion to the faculty's hesitant course, but the prospect will be weighed with care. Both University and School have moved a long way from the administrative practices and procedures of the past. Devolution is, and can be seen as, real. Departments, and not simply heads of departments, have near autonomy. Would a full-time Dean not prejudice that autonomy by becoming involved in co-ordinating departmental planning and development and even in allocating money? Engineering staff, in company with their academic colleagues elsewhere, have relished the democratisation of the campus initiated by Brownlie and will not lightly reverse the trend by giving a full-time Dean authority which might diminish departmental independence. Engineering departments co-operate well, but they are more likely to go their own way than do departments in the Auckland school, over which a full-time Dean presides.

If the Deans were titular leaders of the School, at least for three years, the real leadership came from the departmental heads. Before the move to Ilam the latter were undeniably autocratic; but as the School moved and expanded and the older men retired, a new generation of professors took charge. Their prestige depended increasingly on the quality of the staff they could attract. Fruitful lines of research could result in the acquisition of equipment and research grants, the retention of the best students for research, a high public profile, and a professional reputation. A good head made a good department and the School has been well served by heads of departments since moving to Ilam.

Harry Hopkins was the most successful of the School's leaders after the move. He might not have dominated the School as did Scott, but he was one of the major architects of its transformation over the following two decades. Speculation may be idle, but Hopkins would no doubt

*Sutherland, Alexander James, B. 1939; educ AU (B.E. & M.E. (civil), Caltech (Ph.D.); asst prof Stanford U 1966-67, since 1967 lect., snr lect and reader CU)

have been the ideal choice for a full-time Dean had the post-war College Council done more than toy with the idea in its attempts to recreate something of the School's old *mana*.

Hopkins was, as *The Times* said on his death in 1986, the sort of all-rounder Cecil Rhodes had hoped to develop[6] and he seemed to excel in every activity to which he turned. The obstinate advocate with the stubborn thrust to the jaw was a forthright leader whose academic example, rigorous standards, shrewd assessment of human potential, literacy, scholarship and good fellowship won the respect of those who did not know him intimately and the admiration of those who did. Students were impressed by his athletic bearing and sporting ability — his left-arm cutters were something of a legend in that haven for good cricketers, the President's grade. An initial bluntness obscured a warmth for people and an enthusiasm for his manifold activities that made him a man to be reckoned with in academia, in the community and in the profession, which he led with distinction in 1966-67. Harry Hopkins identified himself with the School of Enginering and the School was fortunate that he did. The Hopkins Lecture, endowed by academics and professional engineers alike, ensures that something of his contribution to School and University will not be forgotten.

His major contribution was to his department. On his arrival, he found it rudderless, badly in need of every resource and with its morale at low level. He left it 27 years later with an international reputation won by researchers whom he had nurtured and encouraged. He was the last in line of the old academic autocrats, but though he ran his own ship he took special pains about selecting the crew. Hopkins acknowledged that it was the staff who made the department,[7] but the quality of his staff was not a matter of good fortune. For a man of action he was remarkably patient about filling vacancies and waited until the right person was available. Some came from the 1700 graduates who passed through the department during his term, but he avoided academic inbreeding by encouraging overseas study for his proteges while judiciously selecting new blood wherever it was desirable. If that inspired loyalty, he too was fiercely loyal to his staff.

Hopkins chafed under the rigidity of salary scales. When he wanted a particular person for his staff, he wanted salary flexibility to get that person. He was well aware that MacElwee's proposal in 1950 for differential salaries in the School was a recipe for endless dissatisfaction, but he lobbied Duncan Mackay, the chairman of the College's Staff and Appointments Committee, soon after his arrival to press for a more liberal policy. Mackay misunderstood. A good deal of staff disaffection existed in the School, he said, and Hopkins should concentrate on building morale if he wanted a strong and stable team. But all Hopkins wanted was a statement on the conditions of appointment that a salary would vary between say £1000 and £1200 according to qualifications and experience. The Council's policy of appointment at the minimum salary regardless of qualifications or experience was not helpful, he said.[8]

There were other reasons for the staff investing confidence in Hopkins. He not only promoted expansion of teaching and research, but practised what he preached. His book *A Span of Bridges* (Newton Abbot, David and Charles, 1970) celebrated his love of bridges and was enjoyed not only by engineers. It was unfortunate that a book about New Zealand bridges remained incomplete on his death. He won the Fulton medal, the profession's premier award; and his abilities at hockey, cricket, golf and bridge as well as his war-time reputation as a bomber pilot made him a popular figure among his students.

But as the department grew and research interests widened, there were signs of restiveness among the staff. A serious illness took the edge from Hopkins's exuberance but it did not immediately suggest greater delegation. There was no palace revolt, but Hopkins, an elder statesman of the University by the seventies, eventually recognised the signs. Three years before Brownlie's appointment, he began to share the wheel with his staff and the result was an easy transition rather than an awkward hiatus when he handed over to Professor Park* whose potential he had observed and noted when he first came to Canterbury. Park did not disappoint him. His research was outstandingly successful, he won a string of prestigious awards, was a good administrator, mixed well with the profession and was a teacher who expected, and received, high standards.

Park was the natural successor to Hopkins. The combination of the Brownlie requirements for consultation in departments and a new broom resulted in a more open yet more cohesive department, despite the wide range of civil engineering activities. It is one of the few departments in the University in which an open door policy is literally observed and the result is an abiding impression of busy yet friendly staff. Park

*Park, Robert B. Fiji 1933; educ Suva Gr.S., CU (M.E.), Bristol U. (Ph.D.); asst eng ChCh Drainage Bd, asst lect, lect. CU 1956-59, lect. Bristol U 1959-65, sen. lect CU 1965-67, rdr 1968, prof. 1968-.

formalised and elaborated the structure of the department with a range of executive and standing committees crowned by a consultative committee, of which he is chairman, and consisting of all the department's professors and readers and an elected staff member. The delegation is real and the Registrar is informed annually of its nature and extent to enable the administration to act on requests within the scope of that delegation. Within the Registry the department has a deserved reputation for efficiency. It also has good communications. As soon as the size of the department's grant is known and a budget determined, full details of all allocations are circulated to staff together with explicit instructions about spending research grants, hiring research assistants and economies in printing and copying. Instructive and informative, the document, issued each December, provides precise guidelines for staff.

If it is a happy department, some of the credit must go also to Tom Paulay* who brought a breath, or rather a gust, of Middle European charm, courtesy and urbanity to the School. After his first degree at Canterbury he spent eight years in a consulting firm in Wellington and was well aware of the profession's requirements. But initially he devoted himself to teaching. As he cycled to Ilam each morning he rehearsed the day's lectures in the heavily-accented English he had taught himself in Hungary from an old textbook; he chose English rather than Russian as something of a comment on the Soviet occupation of his homeland. But it was not long before Hopkins lightened his teaching load and encouraged him to become a student again. He completed a Ph.D. in four years and a half, the swiftest staff Ph.D. known in the faculty, in 1969.

The department gained additional academic leadership with new professorial appointments in the seventies. Ian Wood** continued the teaching of fluid mechanics initiated by Henderson. David Elms*** was appointed to a chair when Hopkins retired. He has broad interests in structures, systems engineering and earthquake engineering and has taken a heavy share of administrative work in the School. His research

*Paulay, Thomas, O.B.E. B. Hungary 1923; educ Hungary, CU(B.E.Ph.D.); struct engr Wgton 1954-61, lect CU 1961-62, s.lect 1963-69, rdr 1970-75, personal chair 1975-.

** Wood, Ian Ruthven ; b 1930; educ CUC(B.E. hons), UNSW (M.E.,Ph.D.); engr Cwealth Works Dept; lect,sen lect, assoc prof UNSW; sen visitor Physics Dept,Cambridge; prof civil engg CU 1971—.

***Elms, David George; b Lond 1934; educ Pembroke Coll, Cambridge (B.A. in mech sciences),M.S.E. and Ph.D. (Princeton);struct engr de Havilland Aircraft Co 1957-60, tchg asst Princeton 1961-63, lect CU 1964, reader 71, prof 78-, Dean 1980-93.

publications include *Linear Elastic Analysis* (B.T. Batsford, London, 1970).

Park and Paulay shared responsibility for establishing the international reputation the department achieved and they and their research students ensured that it was strengthened subsequently. The atmosphere of inquiry and vitality added zest to undergraduate teaching in the department. Distinguished engineers — as many as 20 a year — were attracted from many parts of the world, especially those parts prone to earthquakes, to work in the laboratories and to contribute to weekly research seminars. For 20 years from 1960 it was the largest department too, but in 1980 its enrolment of 262 was overtaken by Electrical Engineering, with 269, and the latter department has drawn steadily away during the eighties, with 303 students in 1985 compared with 246 for Civil. A large proportion of Civil Engineering students were from overseas — 58 per cent of the 159 overseas students in the School in 1985. By comparison, Electrical and Electronic Engineering had 17 per cent, Mechanical Engineering 13 per cent and Chemical and Process Engineering 10 per cent. Overseas students, the majority of them from Malaysia and Singapore, made up 38.5 per cent of the graduate students in the School in 1985.

Growing interest in information processing and perceived employment opportunities in that field were no doubt responsible for the growth in the Department of Electrical ("and Electronic" was added in 1982) Engineering, but the hum of activity generated by lively research groups was not heard till long after the move to Ilam. The department was not initially a prime mover in technological innovation in New Zealand industry and until MacElwee resigned in 1965 electronics did not loom very large in its prescriptions. Pownall was given almost *carte blanche* to find a successor and his choice was Professor L. Kay.* whose research in Britain had resulted in the production of an ultrasonic "torch" to aid the blind and a spate of publicity. Kay, confident and ebullient, was awarded Britain's National Science Achievement Award and he arrived at Ilam like a nor'wester, determined to blow any cobwebs out of the department. The existing course structure, he reported to the Vice-Chancellor, could no longer be maintained. If the output of graduates was to be matched with the growth in technology, a radical change was urgently required. Kay proposed to train engineers in control and communication systems relying on electronics and electronic

*Kay Leslie, B.Durham 1922; educ Durham U, King's Coll, Birmingham U (B.Sc, Ph.D.); R.N.Sci. Service 1948-58; lect. Birmingham U 1958-63; head elec eng Lanchester Coll of Tech 1963-65, prof elec eng CU 1965-86.

instrumentation as well as in power systems, but he warned that a special grant would be required to re-equip the department for undergraduate specialist training in power system, control system and communication system engineering as well as for graduate study and research geared, at least in part, to national needs. If the equipment was not provided, the old pattern would have to be maintained and this would have a serious effect over the next decade.[9]

Kay asked for £143,000 to achieve those aims. Initially the U.G.C. held up its hands. The existing prescriptions, though perhaps biased towards heavy current work, did cover electronics control systems and telecommunications, for which a range of electronic equipment had been provided in the past, said Sir Arthur Nevill.[10] Even so, grants were forthcoming; and publicity about Kay's sonic "spectacles" for the blind suggested that a new era in electronics was just around the corner, as it indeed was. The degree course was restructured to give more emphasis to electronics and development projects, counting as the equivalent of two papers, were introduced into the final year. Kay also initiated a master's course consisting of lectures, research and a project report for graduates seeking advanced training in electronic instrumentation, automation and control. The introduction of new manufacturing industries would require engineers qualified in the new methods. The training of those engineers was an essential prerequisite.[11] Kay called a national conference in 1969 to discuss education in electronics, the justification being:

> Electronics is not merely a technology but a symbol and a symptom. Nothing is more representative of modern industrial progress. The profits a country makes on its exports depend very much on the intelligence, or software, contained in the goods. The value of an export item is in its software content and not its hardware content.[12]

Alas, no Silicon Valley emerged as a result of the talks, but there was progress. Kay threw himself into his own work — though he did not attract many research students — and the seal on his success seemed set when a Christchurch firm began manufacturing the blind aid. His commitment to research did not waver despite his administrative duties and his headship of the department was marked by the establishment of a staff-student liaison committee some time before the Students' Association formally requested departments to consult students. He also stimulated interest by initiating a general studies course in the third professional year.

Kay needed a right-hand man to shoulder some administrative chores to enable him to remain active in research, but initially he was unsuccessful. A British scientist, H. Pursey, was appointed to,but subsequently declined, the chair for personal reasons. Then John Andreae* accepted it only to discover it was too uncomfortable. An engineer philosopher *par excellence*, Andreae deeply regretted spending valuable time on administrative chores — time which could be far better spent on his absorbing research. His integrity was widely known and appreciated and there was only mild surprise when he resigned his chair and asked to be appointed a reader. The administration was aware of his reluctance to leave his laboratory. Once when he was informed that he had been appointed acting head of the department while Kay was on leave, he replied: "I do not accept the position with any pleasure, but will carry it out to the best of my ability."[13] The experience crystallised his determination to resign the chair, a decision he asked the Vice-Chancellor to treat as "the insignficant matter that it is".[14] Phillips replied that he learned of the resignation "with some dismay but with admiration for your sincerity.[15] The drop in his salary was not insignificant, but he returned to his laboratory a happier man and left the chair to a graduate whose M.E. in 1956 was the first to be awarded with distinction. John Bargh** had early experience in computing and his administrative ability was to stand both department and University in good stead. As a scientific officer in the Defence Scientific Corps he worked in the fifties on developing an analogue computer to study the effect of earthquakes on multi-storey buildings and his doctoral research at Cambridge also involved developing a special purpose analogue computer. He arrived as the computer revolution was beginning at Ilam and subsequently played a notable part in helping to formulate University computing plans and policies. In the department he cheerfully accepted wide administrative responsibilities, especially for the supervision of undergraduate studies, finance and equipment. When Kay resigned as

---

*Andreae John Hugh; b 1927; educ Imp. Coll. (B.Sc. hons, D.I.C., Ph.D.; res scientist ICI Ltd and Standard Telephones, sen. lect. elect. eng. CU 1966, apptd to sec. chair 1967, resigned chair 1973, apptd rdr 1973 - .

*Bargh, John Keith; b 1932; educ Piopio DHS, N.Plymouth BHS, CUC (M.E.), Cambridge (Ph.D.); lect elec. eng. CU 1963-66, sen lect 1966, reader 1971, prof. 1973 - .

head of the department in 1982 he ran the whole department with the same quiet efficiency.

The growth in staff numbers as the roll expanded inexorably widened the department's interests as it deepened its investigations. Two professorial appointments in particular in the mid-seventies ensured that two major fields would not only be explored but would also be extended. One of the professors was Richard Bates,* a blunt Yorkshireman and an inspired and inspiring researcher who helped lift the department from academic mediocrity to international attention. Bates may be more scientist than engineer, but he pays as little attention to the distinction as he does to the relative importance of teaching and research. Instead he gets on with academic work at a very high level. He has almost lost count of his publications (some 250 by 1985), his invitations to international conferences and symposia, his election to fellowship of learned societies and professional organisations (seven at last count) and the number of eminent visitors who pass through his laboratories. His graduates, who hold positions of distinction in many parts of the world as well as in New Zealand, are kept informed of his work and of each other's progress by means of a packed annual newsletter, *RHTBOB* (written and distributed by his wife, Philippa).

Bates baffles some staff in other faculties and annoys others. He is impatient with mere competence and he makes occasional outbursts that might be better left unsaid. He recognises them as unworthy subsequently but cannot always desist.[16] He cares little for appearances and his flowing white hair and beard suggest the *guru* from an Arts Faculty rather than an eminent researcher in an electronics laboratory. Undergraduates look on him as eccentric and those who are less enthusiastic than he about their subject might well regard him as a sorcerer, delving into the mysterious and the metaphysical. If he is, he has many apprentices; his research students are not only electrical engineering graduates, but come from other departments and other universities to work with him. He has proved to be an outstanding supervisor and there is a camaraderie among his research group that sustains them all. With research students he claims to be an irritant — the sort of irritant that will produce pearls. He is and he does.

---

*Bates, Richard Heaton Tunstall; b 1929; educ Un Coll, London (B.Sc.eng., D.Sc.); ind. exp. as design and systems engr with Vickers-Armstrong 1952-55, Decca Radar 1955-57, Canadian Westinghouse 1957-60, National Co, Mass. 1960-62, Mitre Corp 1962-65; sen.lect CU 1966, reader 1969, personal chair 1975-.

Bates did not intend to remain in New Zealand initially, but subsequently he has become something of a nationalist — with a small "n" — and has called from time to time for New Zealand institutions to make full use of the talent he and others have nourished instead of continuing with their "pathetic reliance" on overseas experts.[17] And he thought that when staff were elected to fellowship of the Royal Society of New Zealand, the University Council might notice the matter; as he wrote to the Chancellor:

> I remember a few years back that the Council sent special congratulations to Sir Karl Popper on the occasion of his election to the Royal Society of London. It was a very long time since Popper worked here and he is not interested in doing science, but rather in determining what makes scientists tick. (I am one of those philistines who thinks philosophy to be the boringest thing in the world!). I am not trying to knock Popper, who is obviously a great man, nor am I objecting to the congratulations (it was a nice gesture). I am merely trying to put things in perspective.[18]

Bates led his students into fields like image processing and biomedical engineering which would not only have astonished staff a decade before but which brought him — and his students — significant international scientific attention. The appointment of Professor Arrillaga,* a world authority on high voltage direct current transmission, to a new chair in power engineering also brought the department to international attention in a more traditional field. Electrical engineering had been nurtured by the devotion of Bill Bowen** who, almost single-handed, built up facilities in the power field to a standard equivalent to that of the best British universities. His major contribution was the creation of a high tension laboratory where, among other things, a million volt impulse generator was used to simulate the effect of lightning discharges on high voltage transmission equipment. The flash of "lightning" and crack of "thunder" from the generator awed students almost as much as the

*Arrillaga, Josu; b. Spain 1934; educ Bilboa U (B.Sc. elec eng), UMIST (M.Sc. tech and Ph.D.); elec engr. Spain 1956-59, U.K. 1959-62; research UMIST 1962-66, lect Salford U 1967-70, sen lect UMIST 1970-75; prof elec. eng. UC 1975-.

**Bowen, William H.; b 1909; educ Battersea GS, London U (B.Sc. eng. hons); app and design engr Metropolitan-Vickers, 1932-38, elec engr surveyor Lloyds Register 1938-41, engr and chief engr Brush elec. Co 1941-47, sen. lect CU 1947-75.

crowds who thronged the laboratory whenever the School held a conversazione. The laboratory became the only official testing plant for the New Zealand high voltage industry and the Industrial Development Division of the D.S.I.R., formerly of the University, made extensive use of it. An eminently practical engineer, Bowen readily passed his knowledge on to many leading electrical engineers and his expertise in the fields of high voltage and electrical machines was frequently tapped as a consultant.

Arrillaga, whose Spanish accent enlivened his lectures, is closely interested in the New Zealand power system, which took an important lead in high voltage direct current transmission with the opening of the Cook Strait cable, and his research group's publications lead the field. Electrical Engineering divided into several separate research groups, some of which rather lost contact with others. As head of the department Arrillaga is countering that trend and attempting to produce a more homogenous department. A first step, the pooling of professorial secretarial assistance, won immediate approval. And there is a solid sporting camaraderie too. Every lunch-hour a squad of joggers sets out for a taxing work-out. The History Department in its athletic heyday a decade or so ago, was described as "the runningest department in Australasia".[19] The title now surely belongs to the Electrical and Electronic Engineering Department.

When Rastrick retired in 1968 he expressed disappointment at how the sciences, law and accountancy had upstaged engineering in the University.[20] If they had, the department worst affected was his own. Research, as will be seen, burgeoned in the School after the move to Ilam, but the dearth of machinery design and development in New Zealand closed a major research prospect for the department, though several alternative topics were investigated. There was a flurry of interest in the discipline when metal fatigue studies were expanded in the wake of the disastrous crashes of the first Comet jet airliners in the fifties, but mechanical engineering had few of the exciting or "relevant" fields that other departments explored. Heat engines were no longer a mainstay; and though a gas turbine engine was installed, there was not much future for research on that topic in New Zealand. Even so, design, materials and the needs of new industries, especially aluminium and steel, assumed importance. From the late sixties to the mid-seventies significant laboratory equipment was installed, including the first of two electron microscopes, which together with testing machines and analysers, also installed in the late sixties, enabled the material properties of metals to be investigated as never before. It was expensive equipment — almost every piece exceeded the department's annual allocation — but until the mid-seventies many of its major needs were met. The subsequent decline in numbers of research students was disconcerting, but instrumentation and control of machinery and production technology opened interesting prospects for those who looked to the future.

Rastrick was succeeded by a New Zealander whose early research on the problems of the "sound barrier" at the Royal Aircraft Establishment, Farnborough, stimulated subsequent studies at Ilam of the wind, sound and acoustics. Professor D.C. Stevenson* designed the largest wind tunnel in New Zealand for the department and though aircraft design work was limited, the range of uses widened to include measurement of the aerodynamic characteristics of railway engines, rolling stock and road vehicles and the effect of Canterbury's nor'westers on exotic forests as well as large structures. A wooden wind tunnel was built to study air flow patterns over various land forms and a large-scale survey of the winds around New Zealand was undertaken to determine where wind turbines might best be placed if power from the wind was to be contemplated seriously. Increased awareness of the threat of excessive noise to hearing followed systematic noise measurements at airports, in cars and in traffic, at motor-cycle races, pop concerts and factories.

Stevenson led the department from behind, forcing his staff to make the decisions, but he was concerned that young lecturers in particular might, in enthusiasm for their subject, push students too hard. He encouraged the use of teaching aids. For instance Frank Fahy, a senior lecturer, made use of closed circuit TV in the mid-sixties for demonstrating metal specimens under a microscope and considerable use was made of other audio-visual aids in the department.

Mechanical engineering staff have close relations with industry through consulting and testing work—none more so than Professor L.A. Erasmus,** Stevenson's successor as head of the department, whose work

*Stevenson, Dennis Clifford; b 1924; educ Seddon Mem. Tech Coll, AUC (B.Sc & B.E. mech hons), Cranfield (D.C. Ae.), CU (Ph.D.); sci.off.R.A.E. Farnborough, Aero. Research Lab, Melb, sen lect CU 1954-63, sen lect, assoc prof Monash U 1963-68, prof mech eng CU 1969 -

**Erasmus, Leslie Andrew; b 1933; educ Pretoria Tech Coll, UCT (B.Sc. & Ph.D.); F.B.I.s'ship UK; sen. research engr Steel, Peech & Tozer, lect. CU 1965,sen lect 1968, reader 1972, prof. 1978-.

on iron and steel in particular is widely known and respected. Consulting work is encouraged in the School to enable staff to make practical use of their expertise and skills and to extend their own horizons, though care is taken to ensure that the load does not prejudice the prime tasks of teaching and research.

Chemical Engineering (to which "and Processing" was added in 1982) might have been regarded as the School's junior department, but its research was so often a pointer to significant industrial development that it soon gained an enviable reputation for the quality of its investigative and design work as well as a high public profile, especially on the West Coast. Siemon, its founder and first head, laid sound academic foundations for the department and his political skills ensured they did not crumble. He is one of the few persons after whom buildings, or parts of them, at Ilam have been named. In some institutions there is little merit in such honour; at Canterbury, where one needed to be a Rutherford, Hight, Connon, Ngata, Alpers or Jobberns to be so remembered, the honour is real. It goes only to those whose contribution to academic life has been significant.

Siemon's was; but the skilled academic politician had an essential counterweight in Tom Hagyard* whose research on local problems and needs rapidly established the department as a force in regional and national development. At the same time Hagyard was something of an enigma to some of his undergraduates. He was sceptical about the new techniques which were beginning to proliferate in the early sixties and some students found his comments and criticism difficult to accept. He applied rigorous academic standards and demanded the same of students. His special skill consisted of spinning the thread of investigation into a web of applied studies and students were shown gently but firmly how to focus the knowledge and skills they had obtained directly on problems related to the development of chemical industries in New Zealand. The modern design course he established, which remains a key part of the degree under one of his students, Dr W.B. Earl, was oriented to local industry and final year design reports were looked at carefully by Government departments and industries. Hagyard was probably at his best in that curiously intimate relationship between research student and supervisor, but his final year design classes also discovered that he was not only

*Hagyard, Thomas, 1906-1971; b Stockton-on-Tees; educ London U (B.Sc. hons, Ph.D.); chem engr I.C.I., Boots Drug Co, sen lect app. chem. CUC 1950, reader Chem Eng CU 1963-71.

a mentor but a constant goad and guide as well. Many of them look back on Hagyard's tuition with appreciation.

Establishing the course in chemical engineering was valuable enough; expanding it as the School — and New Zealand industry — grew was of key importance. Graduates were in increasing demand in the dairy industry as a host of small factories gave way to a few centralised modern factories; they were needed in pulp and paper manufacturing and in agricultural chemicals and fertiliser production as well as in numerous other fields. The task of overseeing this development fell to Miles Kennedy* whose equable temperament and sweet reason sat easily with his academic ability and relaxed yet decisive leadership. Kennedy was head of the department for 16 years, during which it grew to parity if not in numbers then in influence, with the older departments. He is one of the School's most influential leaders.

Kennedy readily spread his skills. He proved himself as both teacher and researcher and was clearly the best man to head the department at the time. Somehow he developed an uncanny skill for knowing what he could achieve for the department without recourse to "horse trading" and he rapidly won the confidence of the administration. The easy Kennedy style resulted in a happy department He had a flair for selecting the right person for every job and delegated with wisdom. A smooth-running department enabled him to take a leading part in the academic administration of the University** and he was the second Canterbury professor to be elected national president of IPENZ. A democrat in a School that retained an autocratic style, Kennedy prudently shared decision-making and in some measure anticipated the reforms that Brownlie set in train in departments when he became Vice-Chancellor.

Few academics, other than engineers, realise that the School actually consists of five departments; but it must be acknowledged that the fifth, Agricultural Engineering, is with the School rather than of it. It is some 20km of suburbs and farmland from Ilam and its staff are on the Lincoln College payroll. In fact, the latter consideration distances the staff from the School more than the actual distance. The staff teach a B.E. course moderated by IPENZ and are members of the Engineering

*Kennedy, Alister Miles; b Methven 1927; educ Methven DHS, St Andrew's Coll, CU (B.Sc., B.E.) Cambridge (Ph.D.); captain Defence Sci. Corps. 1951-56, sen sci officer Dom.Lab 1956-60, sen lect app chem. Otago U 1960-65, prof chem eng. CU 1965- asst V.C. 1974-77.

**Kennedy was Assistant Vice-Chancellor under Phillips, Deputy-Chairman of the Professorial Board 1980-86 and Deputy Vice-Chancellor 1980-.

Faculty; but though their courses are listed in the Calendar, they are not. And a suggestion that the head of the department, Professor G.T. Ward* might be proposed as Dean of the faculty was swiftly quashed on the ground that he is not a member of the University staff. On the other hand, the department values its association with the Engineering Faculty and its members are not enthusiastic about Lincoln's drive for autonomy.

Agricultural engineering was taught at Lincoln in the thirties as part of various agricultural courses, but it was not until A.W. Riddolls (1909-63) was appointed that a small department was established (in 1947). Riddolls and the Principal, Dr (later Sir) Malcolm Burns (1910-86) pushed strongly for the establishment of an agricultural engineering institute funded by the Ministry of Agriculture, and the opportunity came when deaths from farm tractor accidents began to reach alarming proportions. The New Zealand Agricultural Engineering Institute was established in 1963 to undertake rural engineering testing, research and development and initially to test tractor safety frames. Two years later J.R. Burton** was appointed to be both director and professor of agricultural engineering. Burton, who specialised in water resources, contributed to lectures in civil engineering and there were optional papers in agricultural engineering in the third professional year in mechanical engineering in the early sixties. But Burton wanted a full course in agricultural engineering and in the expansionist mood of the time it was readily accepted. The course began in 1968 and the first students graduated in 1970. Since then about 150 men, and a higher proportion of women than in the Ilam departments, have graduated and filled responsible positions, initially in water resources control but subsequently in a very wide range of farm engineering. The initial emphasis was on the mechanisation of agricultural operations, water conservation, control and irrigation, but the design and layout of woolsheds and milking sheds were also of importance; and as more intensive farming has developed there has been an emphasis on mechanised harvesting, alternative energy sources, especially the wind, solar energy for

---

*Ward, Gerald Templeton b Harrow 1926; educ Glasgow and Durham Us (B.Sc.,mech.eng, & Ph.D.); res off Rouett Inst, engr Malaysia, Singapore 1952-61, prof.eng. McGill U 1961-71, prof ag.eng.Linc.Coll. and director NZAEI 1971-.

**Burton, John Ralph. b Aust 1929; educ Sydney & NSW Us,(B.E. civil); rsch engr U Calif 1961, lect & sen lect UNSW 1954-64, Dir NZAEI & f'dation prof. ag. eng. 1965-70, f'dation prof nat. resources, U New England 1970-.

greenhouses, environmental management and control, food and fibre handling and farm safety.

Students may undertake the intermediate year elsewhere, but they sit most of the first professional year at Ilam, taking an agricultural engineering core subject and a mixture of civil and mechanical engineering subjects.They alternate between Lincoln and Ilam in the second year and do the final year at Lincoln. The department has a teaching staff of 12 (1985) and 23 engineering degree students, but any calculation of staff:student ratios on those indicators will be misleading. In fact, the academic staff of the department have more teaching hours than any other. In addition to the engineering degree course, they also lecture in numerous other Lincoln degree and diploma courses. The B.E. course constitutes less than a third of their teaching load. Technician assistance is more limited than at Ilam. But if some resources are in short supply, both department and institute are housed in spacious, well-designed buildings, the result of a Government promise during the Lincoln College centennial in 1978.

Staff research is at a high level and in 1986 the department had more graduate students than either the Mechanical or Chemical and Process Engineering Departments. The institute undertakes testing and ad hoc research to serve farming needs and departmental staff are able to concentrate on their own special interests in river mechanics, water resource development, agricultural meteorology and wind energy resources, solar energy, aeronautical engineering, especially the design of wind turbines, and intermediate technology . In 1970, no staff member held a higher degree and Burton alone supervised research students; now more than half hold higher degrees and all supervise research.

R.H.T. Bates

R. Park

T. Paulay

H.A.L. Adams, associate professor of Mechanical Engineering, operating the IBM 1620 computer shortly after its installation in the School in 1961.

CHAPTER SIX

A Research Revolution

If there is one activity that most clearly distinguishes the Faculty of Engineering at the University of Canterbury from the National School of Engineering at Canterbury (University) College, it is research. Before the move to Ilam, research in engineering, as in science, was elementary and occasional rather than fundamental and continuous. The new buildings and the growing number of better qualified staff did not of themselves precipitate radical change in what had been a conservative institution, but they were among a number of factors that combined to change the nature of the School and its teaching. Central to this change was a new respect for science and scientific achievement as a result of the remarkable technological advances made during the Second World War and after. Engineering was concerned with the applications of scientific advances and shared something of the scientific limelight, though it was not always advantageous to be included. The achievements of engineers in high technology developments were frequently attributed to scientists,while the failures were as often blamed on engineers. At Canterbury there was also the profound influence of Karl Popper.* whose "nine fruitful years"[1] as a lecturer in philosophy stimulated the intellectual atmosphere immeasurably. The new School at Ilam, if not an architectural gem, provided laboratory space undreamed of in the past and the setting-up grant for the School ensured far better equipment than ever before. An older generation of lecturers gave way, without rancour, to younger men with skills and curiosity who were eager to enhance their own reputations by research and publication. The establishment of University Grants Committee postgraduate scholarships enabled students to

*For an account of Popper's impact on the University, see *A History of the University of Canterbury 1873-1973*, pp 262ff.

contemplate research degrees, especially the Ph.D., and the number of research students expanded rapidly. The first Ph.D. in engineering was conferred in 1952 and there were only 16 research students (eight Ph.D. and eight M.E.) in the total engineering roll of 472 in 1960. By 1965 there were 40 research students, 27 of them working for a Ph.D., in a total roll of 618 and by 1971 there were 100 (62 Ph.D.) in a total roll of 829. The numbers fluctuated subsequently in sympathy with national economic health, but in 1985 there were 104 research students (48 Ph.D.) in the total roll of 895. The scholarships had a direct and profound influence on graduate research.

Research students and the staff had the assistance of a growing army of technicians. Computing enabled research to be undertaken on a scale that would have been impossible in the past. In the *Calendar* for 1960, 12 publications by Engineering staff were listed. For 1986, the number had risen to almost 120. While staff numbers doubled, the number of publications rose tenfold. Perhaps the major reason for the expansion of significant research was the needs of the national economy as the country began to meet and master the backlog of developments delayed by economic depression, war and subsequent austerity. All the engineering departments participated, but the largest, Civil Engineering, was initially at the forefront.

Structural engineering, especially the construction of high rise concrete buildings, began to transform the inner centres of New Zealand's major cities from the late fifties and by the middle of the next decade the concrete industry had grown rapidly. Cement production rose to 840,000 tons a year and when it was combined with aggregate and water to form concrete and poured with steel reinforcement, it had a value of well over $100 million. The size and importance of the industry created a need for research to ensure the efficient and economical use of concrete; and because New Zealand is subject to earthquakes, research was essential to determine the resistance of concrete structures to seismic stress. Building safety, economy and design demanded it.

Hopkins had started research in a small way in the old buildings, but initially his concrete laboratory was a mere six feet by six feet; and though he gained a larger laboratory later, the testing equipment he could command was inadequate. His only recourse was to reduce specimen sizes, but that prejudiced the validity and applicability of published results. Accordingly he sought a 500-ton Universal testing machine in 1955 at a cost equivalent to an unheard of $16,000. The

Council referred the request to the Senate's U.G.C. But Sir Arthur Nevill, to whom the U.G.C. in turn referred the request, recommended that the purchase be deferred at least until the School was at Ilam because the laboratories for structures, metrology , soil mechanics and concrete were "all makeshift, temporary and congested."[2] Hopkins had no time for such reasoning and suggested that the question of balancing equipment against space rested not with the U.G.C. but with those responsible for teaching:

> I would further suggest that the principle that those who have least should be denied anything further on the grounds that they lack the facilities to take advantage of it is a curious one which can only result in our moving in ever-decreasing circles to that inevitable destination which will be known to most members of the Committee.[3]

The College Council's Finance Committee was not amused. It was of the opinion, Logie told Hopkins, that he had not answered the questions raised by Nevill and should re-examine them and comment on each specific point.[4] In the end the U.G.C. did recommend the purchase, but Algie simply refused to provide the money.

The first Ph.D. in civil engineering went in 1952 to John Robinson, who retired in 1986 as Director of Laboratory and Computer Services for the Ministry of Works. Robinson cheerfully acknowledges that there was more design and building of equipment than true research in his doctoral project and there was only very limited staff supervision. Stanton left the School to practise as a consultant and was persuaded to supervise the project occasionally rather, it is suspected, against his better judgment. Difficulties were posed too by the U.N.Z. It was not convinced the School could provide the direction and supervision for an advanced degree and it was only Hulme's initiative in establishing a Professorial Board committee (it included Professors Hugh Parton, Chemistry, and Jobberns, though not Calvert) to be Robinson's guide and mentor that changed its mind. Hopkins, who arrived while Robinson was writing up his experimental work, mostly at home, came along as an observer (though inevitably he took part) to the oral examination, which was held in the living-room of Stanton's home. When it was all over, Stanton shouted for both panel and candidate, a novel but perhaps appropriate gesture for the School's first Ph.D. The first engineer to satisfy the Canterbury requirements for a D.Sc. was Dr R.I. Skinner, head of the Engineering

Seismology section of the D.S.I.R. Physics and Engineering Laboratory, who graduated B.E. in 1957 and D.Sc. in 1977. Only one other engineering D.Sc. was conferred and there are unlikely to be any more. In 1985 the University Council approved the establishment of a D.Eng. degree to be awarded for significant original contributions to engineering,[5] though the U.G.C. initially delayed final approval of the proposal. Between the first two Ph.D.s, the Hopkins policy of patiently waiting for the best men for his staff began to bear impressive fruit. Four were to be appointed to chairs subsequently and two won international renown for their research on the resistance of reinforced concrete structures to earthquakes.

A severe earthquake at Inangahua in 1968 gave real point to the research. A departmental investigation, details of which occupied the entire first issue of the *Canterbury Engineering Journal** showed that most of the damage at Inangahua was to structures not designed according to accepted seismic principles.[6] Practising engineers brought some of the problems designers faced to the attention of Park and Paulay. Despite Hopkins's initial setback, a range of small-scale standard compression and Universal testing machines had been installed in the laboratory for measuring the load-deformation characteristics and strength of concrete, masonry, steel and timber. The concrete strong floors in the laboratory also allowed loading frames to be attached for tests of a variety of structural elements and assemblies. Publication of the results of laboratory investigations stimulated the demands of the profession and Park and Paulay spent a growing proportion of their time preparing for and presenting seminars for practising engineers, especially when there was a radical change in the late sixties to ultimate strength design. Their seminar notes became voluminous and eventually they were expanded into book form. But Park and Paulay's *Reinforced Concrete Structures* (John Wiley, New York 1975, 769pp) is no ordinary textbook. By 1985 it had been reprinted 10 times and translated into Spanish and Chinese. Sales were still rising after a decade. It became the standard international textbook on the topic and its authors have been invited to lecture, give seminars or present papers in all the earthquake-prone countries of the world.

Success inevitably breeds greater success. "P and P", as they (and their book) were known to generations of students, wanted laboratory equipment with a much greater load capacity to test structural elements

*An occasional journal edited by Richard Bates. Most of the five issues published between 1970 and 1975 were devoted to extended reviews of aspects of research in the School.

on a large scale and to simulate severe earthquakes realistically by applying vibrational load to the structure under test. The major laboratory equipment could impose a load of 300 tonnes; Park wanted 1000 tonnes. Confidence in his ability ensured he got it, though the inches-thick correspondence on the appeal for money, which began in 1970, the effects of the rampant inflation of the mid-seventies, and the efforts to bridge the tantalising gap between resources and cost contains a saga of its own. But a DARTEC testing machine costing $275,000 was installed in the laboratory in 1978 after an inevitable problem with ground water had been overcome. It has been in almost constant use since. Involving design and research engineers in structural testing led to further innovative design. Intensive research on large concrete columns suggested that offshore oil rigs, hitherto built of expensive imported steel, might in future be constructed of reinforced concrete.

Much of the research centred on the ductility of reinforced concrete columns and piers, structural walls, and beam-column and slab-column joints subjected to earthquake loading. The behaviour of prestressed concrete structures was also considered. Until the late sixties there were too many uncertainties about the behaviour of tall, reinforced concrete structures during severe earthquakes to permit their construction. Exemplifying the problem was the decision to build the Bank of New Zealand building in Wellington in structural steel. The research cleared those uncertainties away. Analysis and design methods resulting from this work were adopted in New Zealand, American and European structural design codes. Papers by Park, Paulay and their colleagues take major national and international awards and close contact is maintained with other research groups and practising engineers in New Zealand, North America and Europe through membership of numerous technical committees, seminars, refresher courses and visits.

The first Ph.D. for earthquake engineering studies was conferred in 1968 on W.R. Walpole* for work involving prediction of the behaviour of high rise buildings in strong earthquakes. Included among the many Ph.D. students who followed was Nigel Priestley, who joined the staff in 1975 and who has made outstanding research contributions to the behaviour of reinforced masonry and bridge piers subjected to earthquakes and temperature stresses in tanks and bridge decks. His work has won several prestigious awards.

---

*Walpole, Warren Repton. B.1943. Educ CU (B.E. hons, Ph.D.); post-doctoral fellow Imp Coll. Lond 1969; design engr U.K. 1970-72, lect.& sen.lect CU 1973-.

Park did not permit administrative cares to bar him from the laboratory, but he suffered a broken leg while jogging and complications confined him to bed. But the time was not wasted and the result was another textbook *Reinforced Concrete Slabs*, with W.L. Gamble (John Wiley 1980, 618pp). Four or five papers appear in refereed journals each year and in one decade 27 papers by Park and his colleagues or students were presented at international conferences in Delhi, Berkeley, London, Palo Alto, Istanbul, Bucharest, Bogota and Christchurch.

Extensive experimental investigations were also undertaken into timber shell roofs, the methods of joining timber structural elements and the behaviour of the joints, the transverse buckling of steel plate girders and the strength and ductility of bolted steel connections. Analytical investigations included finite element analyses of a range of structures and studies of the dynamic response of buildings, bridges and tanks to earthquake loading.

Hopkins hoped Wood's investigations of Canterbury's braided rivers might also become a major strength of the department; but he found that while money was available for flood repairs, little was freed for the investigation of flood prevention.[7] Even so, good use was made of a fluid mechanics laboratory well equipped for experimental work with a 20 metre long tilting sediment transport flume, several smaller flumes and a stratified flow tank. Staff members Alex Sutherland and Bruce Hunt are closely involved in this work too. Significant and diverse research undertaken includes load transport in gravel rivers, the dispersion of effluent in rivers and from ocean outfalls, air entrainment on large spillways, the problems associated with air removal from pipes carrying liquids, and ground water flow. Staff are closely involved in water resource investigations and environmental inquiries and maintain close liaison with research workers in the field.

Research grants from outside the university system greatly enhanced research in other civil engineering fields too. Alan Williman, who lectured in municipal and road and traffic engineering for 26 years before his retirement in 1980, was a member of the Road Research Unit of the National Roads Board and he proposed the establishment of a test track to investigate the durability of road pavements. The department had taken part in the first major Australasian road experiment, the laying of test sections on the Lyttelton tunnel motorway, but accurate monitoring of traffic over the sections proved difficult. As a result, the unit in 1968 provided the money for a test track at Harewood combining the advantages of laboratory precision and varied road conditions for determining the effect of heavy traffic on paved surfaces. There, two pairs of truck wheels, loaded to the legal limit, revolve round a central pivot at a radius of nine metres, running over up to 12 different road surfaces at a time. Research by Dr P.A. Seddon, a former senior lecturer, into the stabilisation and strengthening of flexible paving led to cost-saving modifications on highways. Some local but inferior aggregates could, the research showed, be used satisfactorily if they were combined with additives like lime or cement. The saving on transport of aggregate from that finding alone encouraged the National Roads Board to spend $330,000 on enclosing and roofing the test track so it may be used at all times for research and undergraduate teaching. New Zealand is believed to have the highest number of lane kilometres of road per head of population of any country, and road research which promises substantial economies and improved paving is likely to attract continued support in future.

Frank Wilson* co-author of four textbooks on waste matter and sewage treatment, led investigations into environmental and public health engineering problems. Rob Davis and John Berrill, in geomechanics, conducted significant research into aspects of the liquefaction potential of foundations during earthquakes.

This multifarious research resulted in the publication of scores of papers, project reports, conference presentations and theses which lined the walls of Park's study . It also brought in more research grants than any other department in the School or the University. In 1971, grants for civil engineering research from outside the University totalled $5800. In 1985 the figure was $488,000. Clearly there is confidence in the profession in the work of a leading University department.

The variety, complexity and excellence of research in the Electrical and Electronic Engineering Department precipitated it to international attention over the last decade and its stature continues to grow. The genius of Richard Bates and the calibre of the graduate students attracted to his research laboratories generated a hum of activity that was almost palpable. The success of his proteges in exacting and exciting positions in many parts of the world as well as in New Zealand bred greater success and the department became an acknowledged world leader in

*Wilson, Francis, B 1940; educ Liverpool U. (B.Sc in chem, B.Sc.(hons) in hydrology, Ph.D. in oceanography; lect civil eng. Strathclyde U 1967 -75, sen lect CU 1976-.

the field of image processing. But if Bates was the dynamo, many others contributed in diverse fields to the high reputation the department has won for its teaching and research.

The success of Leslie Kay in pointing the department far beyond the generation and transmission of electricity opened new avenues that took Canterbury into the mainstream of electronics. The ingenious ultrasonic imaging that led to his developing an ultrasonic aid for the blind is widely known and respected, though the aid did not enjoy the considerable commercial success it promised, partly because costly training was required for its optimum use and partly because modest sales prejudiced a lower retail price. A device for children blind from birth posed a difficult psychological problem: how could children with no innate spatial awareness make use of the aural information about their environment provided by the device? Kay sought the assistance of psychologists. Blind baby monkeys were used in overseas experiments to determine whether they could learn to perceive their world through sound by wearing the device constantly and a Telethon grant enabled spatial awareness training to proceed at the Homai School of the Royal New Zealand Foundation for the Blind at Auckland.

Progress proved slow and Kay turned his attention to adaptations of the principle. They included a cardiophone, a pocket instrument for locating and listening to the positional changes of the four valves of the heart or for monitoring the heart of a foetus. It was designed as a useful adjunct to the stethoscope in family health care centres which had no other sophisticated ultrasonic equipment. Divers demonstrated the value of another adaptation, a head-mounted underwater sonar for locating objects in dark or murky waters. And try as he might, even when fishing became of key importance to many countries, Kay could not get sufficiently firm interest in an underwater fishing sonar to develop it beyond the sea trials undertaken from his own runabout in the Bay of Islands and by Norwegian researchers. He sought research assistance in New Zealand, but was turned down. "No Cash, No Catch" said a heading on a leading article in the *Press* which commented: "It is surprising that a few thousand dollars cannot be found to prove this invention."[8]

Kay was convinced that the path he pursued would yield success. With Peter Gough, now a senior lecturer in the department, whose own research has attracted overseas financial support, he developed a high resolution underwater sonar able to detect and classify objects on the seabed. It would be of value if used with offshore oil rigs and its military potential was investigated in the United States. But he considered his final adaptation was the most promising. With a grant of more than $200,000 from Alex Harvey Industries — the largest research grant ever made to the University by a private compay — he developed the first very high resolution ultrasonic imaging system to operate in air for use by industrial robots.

Kay did not simply lead the way, but hoped his new technology would be put to good use in the industrial world. Reviewing the sonic developments , the *Sensor Review* commented that some had been working for years and added: "The visit to Britain last summer of Professor Leslie Kay . . . was a useful reminder that he began his pioneering work . . . as long as 20 years ago. His impressive demonstration of ultrasonic imaging was a great encouragement to others working in the field."[9]

The sonic aid for the blind was rated as one of two of the most outstanding developments in world-wide medical electronics in the seventies,[10] the other being the brain and body scanners developed by Geoffrey Hounsfield and his research team at E.M.I. With a little luck, computer-assisted tomography, which has revolutionised diagnostic radiology, would also have sprung from an Ilam laboratory for the principle was discovered independently by Bates and a research student, Terry Peters (subsequently the chief image processing expert at the Montreal Neurological Institute) in 1970. Peters asked Bates, who was intending to go fishing one warm, spring evening, whether a large number of conventional X-ray photographs would enable images of cross-sections through the body to be constructed. Bates put off the fishing expedition and they worked out the theory together that evening, but failed to appreciate a crucial advantage of CT image reconstruction in medicine: its remarkable ability to distinguish organs with similar material properties. Doctors at The Princess Margaret Hospital were not therefore specially impressed when Bates and Peters discussed their findings at a seminar a few days later. But in Britain, Hounsfield had already found medical collaborators who by 1970 had recognised the remarkable propensity of computed tomography for enhancing contrast. Bates knew nothing of Hounsfield's success until two years later. He had followed the approved research approach of trying to understand as many general and particular aspects of a problem as possible. But a body-scanner was designed and built by research students for the Radio-Therapy

Department of Christchurch Hospital and was still being used even when later generation scanners were imported.

The research approach was of value, even if one glittering prize eluded the team. Over the next 15 years a stream of research students made numerous contributions to computed tomography practice and then went to research clinics, hospitals and universities in many parts of the world; two to the Mayo Clinic in the United States and one each to the Philips research laboratories in Hamburg, the University of Pennsylvania Hospital, the Royal Hallamshire Hospital at Sheffield, Purdue University and Sydney University. Others undertook medical physics and electronics and computing work in New Zealand. They included Philip Bones, subsequently a senior fellow of the National Heart Foundation; Howell Round, a physics lecturer at Waikato University; and David Wall and Reg Dunlop, senior lecturers in mathematics and mechanical engineering respectively at Canterbury. Bob Jordan and Alan McKinnon had the good fortune to interact with Dr Walter Metcalf, sometime reader in chemistry at Canterbury (and a notable polymath). The former became the electronics expert at the Ruakura Research Station and the latter director of computing and biometrics at Lincoln College. Richard Fright, who was confined to a wheelchair after a diving accident while he was still at school, gained a post-doctoral fellowship to undertake complex high technology biomedical engineering at Christchurch Hospital. David Tan, at 16 the youngest New Zealander ever to gain an honours degree, began research in the laboratory. And Bates's first female doctoral student, Kathryn Garden, was not only the first woman to gain a Ph.D. in electrical engineering, but was also the first woman to be appointed as a lecturer in the department. Awarded a Harkness Fellowship to continue her research at the Mayo Clinic and the University of California, Santa Barbara, in 1986, she impressed researchers there. Dr Richard Robb, professor of biophysics at the Mayo Clinic, reported to Bates that Garden "brings honor to the Harkness Fellowships" and added: "I hope to visit your 'scholar factory' soon".[11]

It was not only biomedical engineering that brought international attention to the research group. Bates was more scientist than engineer, but he ensured that other applications of CT imaging were investigated. Garden scanned sheep to monitor the relationship between their diet and the growth of body fat; the New Zealand Forest Service awarded a contract for an investigation of the use of CT imaging to assess internal defects in logs before milling so they might be sawn in ways to avoid waste; and a process was developed which promised to add significantly to optical astronomy.

When Dr Kenneth Haines, the co-inventor of holographic interferometry, joined the staff in 1971 he established a small optical laboratory and passed some of his expertise on lasers and lenses to Gough, then a doctoral student. Bates was reluctant to take over the laboratory when Haines returned to the United States, but as he delved into speckle interferometry, his curiosity was kindled. He quickly realised that the data processing problems of astronomers, for instance, could be useful training for his graduate students and worthwhile scientific endeavours. The first student to benefit was Dr Peter Napier, subsequently deputy-director of the world's largest radio-telescope in New Mexico, whose doctoral research into holographic techniques for testing the performance of large antennas took him to a leading position among the world's radio engineers. Bates and Gough used a technique they called "speckle holography" to assist astronomers in unravelling stellar secrets. Instead of the conventional single, long photographic exposure of a celestial object they considered a series of short exposures showing the object as jumbled blotches of light; but if the results were visually unattractive, they contained, when processed, much more information than conventional astronomical images. Several graduate students and a lecturer, Fred Cady, contributed to a further development, a technique they called "shift and add", which resulted, after computer processing, in the production of images significantly beyond the capabilities of ground-based astronomical telescopes, as conventionally operated, and suggesting comparison with the results of stellar imaging by telescopes in orbit outside the earth's atmosphere.

Bates and his colleagues covered a broad spectrum. The major impact of his work was undoubtedly in medicine, but he was able to assist a University chemist, Dr Gordon Rodley, to work out the consequences of an alternative model to the famous Watson-Crick double helix model of the DNA molecule on which much of modern biology rests. Another venture was a micro-processor controlled self-teaching aid to musicianship, which was developed by masterate students under Bill Kennedy* who is responsible for much of the hardware and often the software for Bates's projects. The aid was one of the results of a computer-music system which was the central feature of doctoral research by

*Kennedy, William Kennedy. B.1940; educ. CU (B.E. with first-class hons); asst lect 1963, lect 67, sen. lect 75-.

Dr Warren Tucker, subsequently scientific and military adviser at the New Zealand Embassy in Washington. The software was by Dr Martin Lamb, who further developed computer-music systems most successfully at the University of Toronto. Other research students, Drs Alan Jamieson, Geoffrey Burrell and Murray Milner, played key parts in Post Office telecommuniations modernisation in the 1980s and Dr M.J. McDonnell led the image processing group in the Information Technology Division of the D.S.I.R. They constitute a rather devastating response to those who would judge university research only by its immediate relevance; but there was "relevance" too. Short-term benefits as well as greater long-term promise are associated with many of the Bates projects - perhaps none more so than studies of speech analysis and synthesis he and Kennedy recently initiated.

Bob Hodgson* led another image processing group which complemented the work of Bates and his colleagues and which captured industrial attention. In six years the group was awarded more than 20 research grants worth more than a quarter of a million dollars to develop hardware and software for interactive image processing and automatic inspection, or machine vision, for quality control in industry. Hodgson and Cady developed a unique microprocessor-based image analysis laboratory while working on a multi-lens aerial computer camera for use by the D.S.I.R. in resource management studies. That remote sensing work pointed the way to the development of special purpose microcomputers and gave an understanding of how solid state imaging sensors operate. Hodgson was convinced that computers with "eyes" had considerable potential in automatic inspection and saw the potential of microcomputers as the basis of useful interactive image processing systems. A vigorous research programme continues.

Applications of the technology were investigated. They included detecting grass grub damage to pastures from aerial photographs, measurement of canopy coverage in forests and the processing of X-ray images to detect multiple pregnancies in ewes as well as machine measurement of wear in wool carpets. One project which captured wide attention was an attempt to solve one of the perplexing problems posed by an important diversification in primary production. The success of kiwifruit exports is based on quality control; but the large planned increase in production will make effective human inspection of individual fruit

*Hodgson, Robert Martin. B 1946, Yorkshire; educ Goole GS, Bradford U (B.Tech hons), Nottingham U (Ph.D.); lect Hull U 1972-74, lect CU 1975, sen.lect 79-.

impossible. The Kiwifruit Authority, aware of the group's activities, made substantial grants for research into the automatic inspection and grading of kiwifruit at very high speed. Application of similar image processing methods to sawn timber inspection and grading and to a timber recovery process known as finger-jointing could also be of considerable economic significance.

A call for the establishment of a micro-electronics facility in the School was made by D.J. Byers.* on his return from study leave at the Central Electricity Research Laboratories, Surrey in 1972. Without one, he said, the relevance of teaching and research in electronics would be in jeopardy.[12] Byers became discouraged after two futile applications for Government assistance and he focussed his attention on a.c. motor control, subsequently an impressively successful development. His colleague, Neil Edward** discovered that if he wanted a thick film laboratory, he would need to adopt the standard New Zealand approach of the frustrated: to go it alone. He did so, the hard way, designing and building almost everything required but microscopes, wire-bonder and screen printer. A small but effective thick film laboratory became a going concern in 1978. The first substrate was fired in November, the first hybrid micro-electronics course began in the following year and about 100 students subsequently gained valuable "hands on" experience of silicon system design. In 1983 graduate students designed four very large scale integrated (VLSI) circuits, embedding in a single chip several hundred thousand distinguishable circuit elements. The designs were fabricated in Australia and three of the four worked well. The work was ahead of most British universities, partly because of the low-key approach and partly because of the co-operation of Australian colleagues. Establishment of a fabricating "foundry" by the D.S.I.R. at its Physics and Engineering Laboratory encouraged wider acceptance of custom chips and a seminar was given to introduce practising engineers to the principles of NMOS (for negative-type channel metal oxide semiconductor transistor) chip design and fabrication.

Edward, who had considerable technical and industrial experience before joining the academic staff, first demonstrated the value of hybrid

---

*Byers, David John; B 1936; educ Christ's Coll., CUC (B.Sc.) Cran I.T. (M.Sc.); tech signals off. RNZAF 1958-63; lect CU 1964, sen lect 1970-.

**Edward, Lionel Neil Milne, B Greymouth 1932; educ ChCh West HS Portsmouth Coll of Tech., Southampton U. (B.Sc.hons); RNZN radio electrician, electronic engr. G.E.C. Portsmouth, sen.res. engr. Plessey Co; lect CU 1969, sen lect 1973-.

circuitry by designing a tiny, specialised transmitter enclosed in a tranquilliser dart for use in live deer recovery. Another successful design was an a.c. waveform analyser for power supply authorities. Both designs were exploited commercially.

The use of high technology electronics designed in the department had a significant impact in the field of heavy engineering too. Byers applied his knowledge and experience to design an a.c. motor controller which enabled simple and reliable a.c. induction motors to be varied in speed in the same way as the less reliable d.c. motor. The advantage is that because the motor's speed may be matched to the load, the motor runs more efficiently and more economically. Industry rapidly recognised their value and the Napier firm of Industrial Electronics and Automation Ltd manufactured the controller for New Zealand and overseas sale, especially for controlling large pumps.

If the Byers invention was the most successful commercial development of a research project, the work of John Andreae into artificial intelligence may be of far greater significance in future. Andreae concentrated on the development of general purpose learning systems which do not have real world knowledge built into their design and which therefore must acquire it through experience. He described in his book *Thinking with the Teachable Machine* (Academic Press, 1978) how a system he called PURR-PUSS, simulated in a computer, could be taught simple tasks like remembering nursery rhymes, counting objects or drawing simple pictures. Andreae believes the system is similar to a human brain and will need to be taken through all the stages of baby and child development described by Piaget before significant advances are made. The education of PURR-PUSS is therefore likely to be protracted. Considerable time has been spent also in a study of engineering robots to enable his creation to be provided with vision and arms. Andreae had little financial support - the Defence Department made limited research grants — and few research students were attracted. Indeed, in the field of artificial intelligence he ploughed a lone furrow. But there were occasional encouragements. One was the publication of his book and another was an international conference in 1986 on future advances in computing. It was held at Ilam and 30 experts from North America and Australia working principally on performance-oriented systems, acknowledged the international interest in Andreae's work. Few major developments have yet occurred in artificial intelligence and they saw promise in Andreae's approach.[13]

Electricity is the most expensive and the most vulnerable of all the large systems like water reticulation, sewerage and communications that are a major but somewhat overlooked contribution by engineers to modern society. Yet users seem remarkably unaware of the political, managerial and technical decisions that are constantly required to maintain a reliable, economic supply. A breakdown or unwise scheduling of hydro and thermal generation soon costs the country millions of dollars. The techniques used to control and optimise a complex power system are therefore extremely sophisticated. Indeed the computer requirements to optimise the operation of a large power system are just as demanding as those required to send a man to the moon.

The appointment of Josu Arrillaga enabled the School to make a major contribution to electrical power systems research. He had been in charge of the power system group at UMIST, the strongest group in Britain, and he felt quite at home in his new environment largely because New Zealand, with the Cook Strait cable, had taken a lead in h.v.d.c. power transmission, his main research interest. In 1976 a newly-constituted power system group, which included Dr C.P. Arnold, also from Manchester, began work on a new generation of computer programs to incorporate h.v.d.c. links within conventional power system analysis. Some 13 Ph.D. and numerous M.E. theses have been produced since then and the group has won a high international reputation in h.v.d.c. and power system harmonics. Proof of its success were the substantial and regular contributions to the principal international and professional journals and conferences. An annotated bibliography of international research on transmission published by the American Institute of Electrical and Electronic Engineers and covering research since 1970 showed the Canterbury group had a record number of entries.[14] Three specialist books,* each a leading document internationally, have been published and much of the software developed is in use by some of the largest electrical manufacturers in Europe as well as by the Electricity Division of the New Zealand Ministry of Energy.

But despite the intrinsic interest and obvious importance of the topic, graduate students in recent years were more likely to be attracted to computer and information technology, the effect of which is clearly

*The books are *Computer Modelling of Electrical Power Systems,* John Wiley, London 1983; *High Voltage Direct Current Transmission,* Institution of Electrical Engineers 1983; and *Power System Harmonics,* John Wiley (London) 1985. Arrillaga's co-authors were C.P. Arnold, B.J. Harker, D.A. Bradley and P.S. Bodger, now a senior lecturer.

visible in the field of consumer electronics. The power systems group found it difficult to maintain a high international profile with New Zealand students and relied increasingly on overseas students to undertake research.

Thirty years before, the handful of research students in the department had also been attracted to the more spectacular side of the discipline. Bernie Withers,* one of the last lecturers to be appointed without formal academic qualifications, made a considerable public impact — and a little niche in history — by making the first television transmission in New Zealand. Withers trained a group of students, including W.S. Strong, R.D. King, D.G. Reid, B. Gordine and L. Gruener, in vacuum tube technology and decided that building a transmitter would be an interesting way of learning to use the tubes. Station ZL3XT in the electronics laboratory began test transmission in mid-1952. Its signal penetrated to deepest Fendalton and Riccarton and to Linwood in the east and was picked up by about 30 (mostly home-built) receivers. In May 1953, Withers decided to give the annual conference of the New Zealand Radio Traders' Federation a foretaste (as the *Press* put it) of what they and their customers might see on their receiver screens in future.[15]

This first telecast was a technical success even if the entertainment was far from compelling. King read a story on camera and then answered a telephone call from Withers in the University Hall across the quadrangle. "I guess you are getting pretty hot there," were his rather less than historic words. King agreed that it was "pretty warm." Their voices, the *Press* said, seemed to come from the screen of the receiver, but they actually came from speakers at the back of the set.[16]

It is not surprising that well-equipped graduates and valuable research findings should flow from the department at Ilam. It has had an average of 36 research students for two decades and together with the academic staff of 21, they must constitute the largest research group in New Zealand. But their interest, like that of their counterparts throughout the world, has switched from traditional electrical engineering to computer and information technology. It is not so much the techniques as their applications which attract the attention of the news media and, in turn, students. That polarisation makes it difficult for the power systems group to maintain its high international profile.

The problem was worse in the Mechanical Engineering Department,

*Withers, Bernard Tyndall, 1906-83. Educ. Waitaki and Timaru BHS; AMIEE; radio engg US and NZ, designed radar for armed services in Pacific; lect and sen.lect. CU 1946-70.

where the slow but steady growth in undergraduate enrolment was not matched by graduate study. The academic staff, more eager than their predecessors to undertake significant research, were frustrated by a growing shortage of research students, not because of attraction to fashionable topics but rather because of the yawning gap between the value of graduate scholarships — a maximum of $8000 — and the salaries normally offered to new mechanical engineering graduates of $20,000 or more. In 1965 the department had five research students. Numbers rose to a peak of 31 in 1975 and declined steadily to a low of nine a decade later. Many of them were overseas students; but staff who felt any disquiet about the proportion needed only to remind themselves that New Zealand students too went overseas to complete research degrees.

The electronics research aimed at enhancing industrial robots was complemented in the Mechanical Engineering Department by Professor H. McCallion,* an engaging Englishman who switched from a chair in mechanical engineering to one in production technology.** In addition to wide-ranging management studies, he and several research students investigated the difficult problem of automatically fitting a peg into a hole with clearances of hundredths of a millimetre. It is a difficult task even for a skilled person; the peg frequently jams because of the minimal clearance. Initially the researchers attempted to simulate human touch, with four sensors feeding information about pressure changes back to a computer, which corrected misalignments. But theoretical analysis suggested that a passive compliant device, which simply allowed misalignments to be corrected without computer direction, would be a more rewarding approach. It was. Five graduate students under McCallion's supervision developed what looks like a set of springs on legs allowing a component to be pushed so that misalignments are gradually reduced until the alignment is perfect and the peg drops into the hole.

Assembly workers are seldom able to fit a shaft with a diameter of 50mm into a hole with a diameter of 50.05mm without the shaft jamming. Under test with the D.S.I.R.'s Unimate robot, the device worked perfectly, completing the task 120 times with no jamming. Close

*McCallion, Henry, b 1927; educ London U (B.Sc,Eng.), Nottingham U (Ph.D. D.Sci.); eng.training 1943-50, Liverpool, grad. study 1950-53, lect, snr lect, rdr, Nottingham U 1953-70, prof mech eng CU 1970-73, apptd to chair prdn tech 1973-.

**The chair was endowed in 1972 by an anonymous benefactor to mark the University's centennial. centennial.)

overseas interest was followed by continuing evaluation of the device in Europe.

Proof that metallurgical analysis and inquiry are not necessarily tedious and unrewarding was provided by Erasmus with numerous accident investigations, including one ingenious piece of metallurgical detection work that resulted in the withdrawal of an aircraft accident report and the award to him of the Wigram memorial medal of the New Zealand division of the Royal Aeronautical Society. In 1961 a Bay of Plenty Airways Aero Commander crashed on Mt Ruapehu, killing the pilot and his five passengers. The air accident report blamed the pilot, but his colleagues and friends were not satisfied with the finding and they eventually sought expert assistance.

Erasmus thought that if the pilot had flown too low (as the report suggested) and a spinning propeller had hit a rock outcrop at the summit, rock fragments would have been embedded in the metal. Instead he found traces of aluminium alloy in the bent propeller tips, indicating that the tips had struck the fuselage. Further investigation revealed that a fatigue crack in the wing spar caps had allowed the starboard wing to swing forward, forcing the propeller into the fuselage and causing the wing and engine to fall off. Twenty-one years after the crash, the first accident report was officially withdrawn, and the fatigue crack finding accepted. But the pilot was not exonerated, though Erasmus is convinced that the wing failure was inevitable. A further experiment proved that the propeller tips bend forward on impact with the fuselage.

In addition to continuing research on metal fatigue, close attention has been given to atmospheric corrosion of aluminium, even in the Antarctic, to the explosive joining of steel reinforcing rods, the instrumentation and control of machinery and advanced manufacturing technology.

The four departments at Ilam seldom co-operate in their research, but a notable exception — perhaps a pointer to the future — has been the involvement of the Mechanical Engineering Department in recent energy-related research. Installation of a "rolling road" chassis dynamometer and allied equipment under a contract with the Liquid Fuels Trust Board enabled the efficiency and economy of various fuels, some from nearby laboratories, to be assessed and vehicles to be tested. A rigorous investigation of domestic energy use was made as a first step in its reduction. An elaborate laboratory installation is measuring domestic heating and ventilating, how much heat from waste water can be transferred to incoming cold water and the effect of solar water-heating. This research complemented the development by Professor A.G. Williamson* of a solar collector for domestic water heating based on the principle of a heat pipe and enabling steel rather than expensive copper to be used in its manufacture. The collector was subsequently manufactured commercially.

But the most spectacular example of co-operation was a battery-powered electric car, a special project of David Byers and Richard Harman, a senior lecturer in mechanical engineering. Curiously, none of the funds that became available in some abundance for energy research in the seventies were made available for the project and when the Minister of Energy, Mr Birch, visited the School in 1979 to observe, with some satisfaction, the devotion of the staff to energy research, he bluntly told Byers that the Government was not interested in electric vehicles. Starved of money, the project all but stalled and development has been slow. Yet the vehicle has promise, especially if (as seems inevitable) oil costs rise in future. It has several unique features, including novel static inverter-powered a.c. induction motor drives, it will be silent (eventually) and non-polluting. The first version of the vehicle attracted an overflow attendance when it was featured in the first of a series of public lectures on University research in 1977. There were even complaints about over-crowding. Certainly the acting Vice-Chancellor had to sit on the stairs.[17] The second version of the vehicle, looking like a conventional car, with its 20 batteries discreetly out of sight, attracts rather less attention. Indeed the news media seemed more interested in Tricanter, a pedal-powered tricycle that its reclining rider hoped to push to 80 kmh.

Before research was fully recognised as an essential function of an engineering department, Hagyard was encouraging students into research laboratories and a significant graduate school was established in the Chemical Engineering Department. It was significant because while students looked into fundamentals, they were (somewhat to their surprise) soon closely engaged in applied studies, particularly the investigation of possible new chemical industries. Hagyard had a keen eye for potential developments and became an honorary — and honoured — West Coaster as he and his students investigated the prospects of establishing a calcium carbide industry, the production of titanium dioxide from Westland beach

*Williamson, Arthur G. b 1932; educ CUC (M.Sc.hons), Reading U (Ph.D.); post.doct. research UCLA; sen.lect. chem Otago U, sen lect chem eng CU 1967, prof 1974-.

ilmenite sands or the range of chemicals that might profitably be made by extracting turpene from radiata pine. Long before the Bluff aluminium smelter was established he investigated the problems associated with the production of anode carbon from low-ash Stockton coal in the Buller field. A decade later, when the smelter was fast becoming fact, that research was swiftly resurrected and Government funds were provided, on a scale that would have astonished Hagyard, for studies of the coking, grinding and briquetting required, a computer simulation of an electric calcining furnace and design of the optimum process conditions. Similarly, he extended graduate studies of the viscosity of fluidised systems to the possibility of fluidising a channel in the seabed to prevent the build-up of sand bars at the mouths of the two river ports on his beloved West Coast. That sort of work, far ahead of others, was followed with enormous interest "over the hill". Westland faced hard times as its coalmines closed and new industries were eagerly investigated.

In the end, Hagyard received little official encouragement, though it must be acknowledged that his unorthodox adaptations of equipment in making proposals provided loopholes for criticism. Perhaps the close official analysis of the bathwater obscured the value of the baby and both were thrown out.*

Kennedy's research interest was ion exchange, but in Christchurch he soon became involved in air pollution studies. The city, with a temperature inversion problem in winter and householders who treasured open fires, was — and remains — an ideal laboratory for research on a problem that threatened to become a serious health hazard. His modelling of urban air pollution helped with predicting the conditions for and the onset of smog. Health and local authorities took note and public acceptance of measures to reduce urban smoke emission is slowly increasing.

But Kennedy's major contribution was undoubtedly his confident management of the department. The times might have been propitious, but the well-qualified staff and sophisticated equipment did not arrive by accident in the five-storey block that distinguished the department from the rest of the School. While some staff worked on short-term

---

*Hagyard worked hard on proposals for the exploitation of Kapuni gas instead of using it for electricity generation and suggested a process that would have resulted in the production of methane and LPG, a portion of which he proposed should be shipped by barge to provide the South Island with a share of a national asset. No document can be found to confirm his story that the official response, while expressing polite interest, suggested the proposal was not very practicable because there was unlikely to be any market for LPG.

but essential research projects, investigations continued on the thermodynamic properties of liquid and gas mixtures, coal utilisation, process control and drying.

When he came to the University from Britain in 1962, Roger Keey* was closely interested in mass transfer, but he was strongly influenced by his environment. A survey of mass transfer that he made in 1965 showed that drying processes — milk, wool, paper and foods -were the important applications in New Zealand and he switched his research emphasis. In short order he became an authority on industrial drying and published two texbooks, one of which was translated into Russian, as well as numerous papers providing a theoretical foundation for drying processes and an appreciation of the problems underlying the design and operation of drying equipment.

International recognition soon followed. Keey conducted courses on drying for Swiss, British and Canadian engineers, and gave research lectures and seminars at universities in Europe and North America and spent study leaves as a research associate at Harwell and at the Institute of Thermal Process Engineering, Karlsruhe. At home he won significant research grants, particularly for investigations into spray drying milk products, an important sector of the national economy. His subsequent work in hazard analysis stemmed from his consulting activities and was typical of the spread of the interests. He turned historian to produce *To Him Be the Glory* (St Paul's Trinity — Pacific Presbyterian Church, 1975), a pleasing account of Mountfort's quaint and dignified little architectural gem, once Trinity Congregational Church, in Worcester Street; he devoted himself to the Canterbury Workers' Educational Association; and his Meccano model of the Maui offshore oil rig, which attracted close Ministerial attention at the Chemeca 83 conference at Ilam indicated that he might have been equally at home in another branch of engineering.

The swift and effective manner in which University research can flourish when money is available was demonstrated by the department and the School in the wake of the oil crisis of the mid-seventies. As the proportion of hard-earned foreign exchange spent on oil imports mounted, substantial research grants were made by the New Zealand Energy Research and Development Committee, one of the most successful creations of the Kirk Government, for a thorough investigation of

*Keey, Roger Brian; b 1934; educ Birmingham U (B.Sc.hons, Ph.D.); ind. exp with Distillers Co. 1957-61, lect CU 1962, sen. lect 1965, reader 1971, personal chair 1978-.

alternative liquid fuels. Dr Earl, Dr Earl Graham and a number of research fellows investigated in particular the use of methanol, readily manufactured from the abundant natural gas New Zealand had tapped, as an alternative fuel for the country's vehicle fleet. Staff, or rather their cars, were the guinea pigs in research which indicated that a mixture of 20 per cent methanol and 80 per cent petrol could be used satisfactorily for most cars with no modification of the engines while a mixture with those proportions reversed could readily be used in engines modified at a cost of up to $1000. The use of methanol in small two-stroke and large diesel engines was also investigated and the potential savings in foreign exchange were shown to be extremely significant. In the end, the Government decided to manufacture methanol from gas reserves, but an additional process transforming it into synthetic petrol was finally chosen, at considerable cost, to reduce the country's dependence on imported oil.

The Minister of Energy, his Under-Secretary and members of the Opposition came to Ilam to see at first hand the wide range of energy research. They were specially interested in the investigations into the use of methanol and in Earl's comparative survey of other liquid fuel options and the prospects of future fuels from the biomass. They heard from Ian Gilmour, a senior lecturer in the department, about a unique fluid-bed furnace he had devised as a solution to waste disposal and energy problems; from John Peet, also a senior lecturer, on his energy analysis of all the country's goods and services; how solar heating devices were being developed; alternative diesel fuel; reduction of domestic energy consumption; and several other projects which showed promise of helping to avert an energy crisis. It was an impressive demonstration by all departments of the School of how closely university research is related to the social, industrial and economic wellbeing of the country and of how rapid the response can be when adequate research grants are provided.

Construction of the new School begins at Ilam in 1958

The School in 1986 with the University buildings spread over the vacant land prominent in the picture above.

CHAPTER SEVEN

Hard Work, Hard Play

It is difficult to measure the effectiveness of what is, after all, the University's prime task of teaching, but there is no doubt that the School deserves the high reputation it has achieved. Its graduates, as will be seen, have made a significant contribution to the well-being of New Zealand society and employers, including recruiting representatives of overseas companies, continue to seek them out. That is scarcely surprising, for the B.E. is said to be of a higher standard than in the United States.[1] Certainly New Zealanders undertaking graduate study in North America are as well, if not better, prepared than local graduates.[2] It has also been estimated that Engineering School entrants exposed to the National American SAT admission examination would score about 1150 marks and those entering the first professional year about 1250 of a possible 1600. Those marks would place them below entrants to the Massachusetts and California Institutes of Technology, who have a median score of 1400, and four other American engineering schools, but they would be ahead of most others.[3]

It would be surprising if the School did not rate highly, for students are selected for entry by the intermediate examination. About 10 per cent are qualified by outstanding performance in scholarship or bursary examinations to skip the intermediate year and students with good passes in the New Zealand Certificate of Engineering examinations are admitted directly to the second professional year. But for the others, the intermediate year must be mastered before entry. Numerous studies have shown that students with high entry qualifications fare best at the University and engineering students are no exception. In 1986, more than 40 per cent of all the Junior Scholars in the University were engineering students. More than 50 per cent of the School's students (compared with 30 per cent for the Science Faculty) entered the University from the national lists of Junior Scholars or "A" bursary holders. A high proportion of distinguished academic records — second in number, but not in quality, to Science — is presented by engineering students for the award of Senior or U.G.C. Scholarships.

In 1985, the overall examination pass-rate of engineering students typically exceeded 90 per cent. In the last two decades only the School of Forestry has recorded a higher rate. The graduation rate for students who win admission to the School is correspondingly higher than other faculties too. The most recent measure of engineering students' progress to graduation after entering the School, shows a graduation rate of 85 per cent with a further 5 per cent still enrolled. About 90 per cent of those admitted to the School therefore graduate.

Thirty Canterbury students have been awarded Rhodes Scholarships up to 1985; seven have been engineers: A.G. Bogle (1937); G.D.G. Dalton (1939); J.W. Ridley, who was subsequently an M.P. (1946); D.L. Schultz (1951); W.S. Wakelin (1964) , R.E. McConnell (1970); and D.McG. Elder (1980). Timothy Langrish, a chemical engineer, won the 1985 Royal Society Rutherford Scholarship, only one of which is awarded throughout the Commonwealth each year.

The School has been a magnet for overseas students, many of whom now practise engineering in Malaysia, Indonesia, Vietnam, Singapore and other countries. These students add new dimensions to life in the School and, in growing measure, to Christchurch. If the first trickle in the early sixties felt a little lost and bewildered, it was an engineering student, Trinh Khanh Tuoc, from Vietnam, who helped change attitudes by writing *Our New Zealand Experience* (Caxton Press, 1968), a book which helped the newcomers to understand New Zealand and its ways — and local students to see their guests in a new light and themselves in a mirror which was not always flattering. But the trickle soon became a flood: in 1965, there were only 30 overseas students in the 231 who enrolled in the engineering intermediate year. By 1970 there were 202 overseas students in an enrolment of 390 competing for 200 places available in the School in the following year. Alarm bells began to ring. A quota system was established and in 1986 there were 31 overseas students in the intermediate year enrolment of 238. Despite restriction on entry, overseas students comprise over 18 per cent of the total School roll, constitute about 50 per cent of postgraduate enrolments and they outnumber local women students by about four to one.

Perceptive staff members recognise that women constitute the major pool of quality recruitment for engineering in the future, but the School has not been signally successful in tapping that pool, in spite of a joint campaign with Auckland to encourage high school girls and their careers advisers to give serious thought to engineering as a career. As a small but growing number of female students has shown, there is no intrinsic reason why women should not become good engineers. But in the 20 years since Leslie Robertson graduated B.E. in electrical engineering in 1965* there were only 65 other women graduates up to 1985. Women constituted 1 per cent of the roll in 1970, 2 per cent in 1975, nearly 3 per cent in 1980 and more than 4 per cent in 1985. But the number in 1986 was only 56; and even if the upward trend does continue, the change seems certain to be evolutionary rather than revolutionary. As Professor W.L. Jones, a former Dean of Science, pointed out, if women were to account for as much as 25 per cent of graduates in future the numbers of women students would need to increase by a full order of magnitude.[4]

The School faced problems in encouraging the enrolment of women. Despite considerable publicity for the pioneers, the paucity of precedent results in a Catch-22 of few role models and few enrolments as a result. Fewer girls than boys retain mathematics in the senior school[5] and some careers advisers discouraged those who do make the right educational choices from considering engineering as a career.[6] And if the social climate has changed, it is still disconcerting for a young woman to enter a school in which only 3 per cent of the teaching staff and 4 per cent of the students are of the same sex as she. For some, the ear becomes attuned to hints of patronage, prejudice and chauvinism in lecture or laboratory and a few women graduates report detecting them.** But it is clear from responses to a survey of women graduates that such treatment is the exception rather than the rule. However, most women acknowledged that they felt conspicuous and as a result were not completely comfortable in the School; but they thought that

*Leslie Robertson was not the first woman to qualify as an engineer through the School. That honour belongs to Mrs Pat McCook, who took lectures in electrical engineering, enabling her to pass examinations for entry to the Institution of Electrical Engineers and thus to qualify for entry to the profession in 1955. She subsequently joined the I.D.D. staff.

*Examples, from an admittedly small survey, include a lecturer who persistently addressed his class with "Good morning gentlemen", though not all its members were; one who, according to the informant, "gave off vibes that women were not really meant for engineering nor was engineering meant for women"; and another who persisted in "irritating" sexist jokes.

generally the staff encouraged them more than male students. Comments included:

> Maybe we were over-protected. Out in the real world I have experienced a few problems. There was always that slight challenge from my male (student) colleagues to do well to prove myself as a female (but not to do too well and better them.)
>
> I found the staff and fellow students both friendly and helpful though they usually underestimated the female students, which can be slightly irritating at times. (A Malaysian view).
>
> My advice to girls contemplating engineering is to persevere.
> It is certainly a hard road to travel. Your failures will always be disasters and your successes miracles . . . Women do not have to be brilliant scholars to be engineers.
>
> The School has a reputation for being a place of hard work and I too was swamped at times by 9-1 lectures and afternoon laboratories. It is difficult to assess the material taught by the School as there is so much to be covered. However, I have to conclude it must have been adequate preparation for the "real world" as I have been able to cope with my job as an engineer.
>
> Engineering School is quite separate from the rest of the University — quite self-contained, with its own library and cafeteria even. This is a pity as perhaps more contact with other faculties would help remove the School's image of being a boring, hard-working, chauvinistic male bastion. It might then be a more appealing option for women.[7]

If practically all the staff were sensitive to their needs, the first wave of women students showed courage and determination to survive three years of being surrounded in lecture room, laboratory and library by males. They discovered real prejudice and discrimination when they went into industry to meet practical work requirements and finally to gain employment.[8]

The absence of a significant female enrolment not only slowed progress towards sex equality in an important profession, but affected social life in the School. Engineering students have long had a reputation for boisterous behaviour, as this verse (sung to the tune of *The British Grenadiers*) from Diploma Day at the turn of the century suggests:

Some talk of Deans and Doctors and some of LL.B.s,
And grow inept at grinding for their dull and dead degrees,
But for broad-brimmed jocularity, there's none that can compare,
With the rowdy dowdy boys of tow, the College Engineers.[9]

The successful experiment some 70 years later of mixing the sexes in halls of residence at Ilam suggests that more women in the School might well restrain excessive exuberance by "rowdy dowdy boys" in their extra-curricular activities. Women engineers will no doubt play many roles during their lives. In addition to being students initially they may also head a kind of *mission civilisatrice* in the School, as did their sisters in University, Rochester and Rutherford Halls.

Students are usually judged by the excesses of the few and engineers are no exception. It would be remarkable if hundreds of young men did not kick over some traces after long, lonely hours of study and examination. Excessive drinking is not unknown among other faculties on occasion; nor is it uncommon beyond the University among that age group. Yet students in general and engineering students in particular provoke hostile public reactions at end-of-term celebrations and the sins of an unwise minority are readily visited on the majority.

For many years, engineering students were discreet about social behaviour. Ensoc, which had been established in 1897 with A.R. Craddock (1877-1949), as first president, is the most active of all student societies, and continues to organise an annual ball and a dinner, both conducted, it was carefully explained in 1950, with "the dignity and behaviour befitting the profession." Ensoc organised a smoke concert in the late forties, subsequently a staff-student social, and an annual rugby match against mining and metallurgy students at Otago University. Inevitably there was a drinking contest too: it was described as a "social function" at first,[11] but by the mid-sixties it emerged in its true colours as a drinking horn. Ensoc was beaten at rugby in Dunedin "although a well-organised 'bash' with the physiotherapists on Saturday night did more than enough to compensate for the sporting frustration."[12] One change in the social programme had a curious result. Ensoc decided in 1966 to change the venue of the staff-student social to the Horticultural Hall:

> thus endeavouring to avoid the noisy swill atmosphere associated with previous socials in the University Hall . . . the net result of this move was that Ensoc was billed with the cost of repolishing the floor of the Horticultural Hall.[13]

Small wonder that G.B. Page, chairman of the Ensoc social committee, should comment that there was one thing he had learned from the job: "Never underestimate the drinking capacity of engineers."[14] That capacity was sometimes overestimated at a subsequent Ensoc affair, the annual tug of war at the end of lectures.

Ensoc did more than promote beery hilarity. There were less lubricated sporting occasions, like the staff-student hockey and cricket matches promoted and enjoyed by Hopkins. There were at least ten evening meetings a year at which formal papers were presented, visits to factories, orientation for freshers and service at other public occasions organised by the School, such as the three-yearly conversazione. Ensoc also published a selection of student papers as Proceedings, but in 1948, with a little help from the College Council and Students' Association, it published the first of 22 issues of the *Student Engineer*, a rather unlikely candidate for literary longevity. Its first issue, of 162pp, contained a dozen technical papers, an editorial suggesting that social studies had no place in the engineering curriculum, notes on the School, the staff and graduates and a report on the surveying camp — it was held at various places, such as "(un)Godley" Head and Burnham military camp — written with all the breathless excitement of a school magazine contributor.

The "new look", a fashion which decreed that women's skirts should reach almost to the ankles, spread through the College as well as the city, but its only manifestation in the School was a new Ensoc constitution providing for greater continuity of policy, the election of graduates to membership and more meaningful activities.[15] Possibly the high point for Ensoc was in 1953 when it officially became the student chapter of the Institution of Engineers (subsequently IPENZ) gained scholarships from industry, won strong support for all its social and technical occasions and managed to enrol 238 of the 256 students. It also had numerous entries for its Templin Scroll and Templin Prize awards for the best student technical paper.*

Rolleston House, the men's student hostel — it could scarcely be described as a hall of residence — played an important part in Ensoc's

*The awards were made by J.R. Templin (1881-1961), who graduated M. E. from Ohio State University and who came to Christchurch in 1905 as an employee of the General Electric Co. to supervise the electrification of the tramway system. It was not his sole interest. He married the daughter of the city's fifth Mayor, took her back to the United States and returned in 1910 to work as an electrical engineer on the West Coast goldfields. Subsequently he went into partnership with H.F. Toogood and they were consultants to numerous power boards. Templin was an "outside" member of the Engineering Faculty for 14 years.

fortunes. From its opening just after the First World War, it attracted students from all parts of the country and many of them were engineers. Indeed more than 300 engineering students are shown on the *House List* and some took a leading part in student life. One of them was J.B.C. Taylor* who was secretary of R.H. in 1939 and president in 1940 as well as being president of Ensoc and of the Students' Association. He was also NZUSA president 1943-46. The engineering faction exerted a profound influence on student activity. In the spacious backyard of R.H., as the large, old timber houses were affectionately known, students used their elementary construction skills to produce elaborate floats for Procesh. There the late (and unlamented) haka party in grass skirts, hobnailed boots and rugby jerseys, practised their insensitive parody of a haka before taking to the streets during Capping Week and demanding money for charity by (mild) menaces.**

And there some of the best student stunts were planned. Only engineers, it was said, could have strung clotheslines containing clothing (some undoubtedly feminine and purloined from Helen Connon Hall) across Worcester street, penned a mob of sheep in the main College quadrangle overnight, tinkered with the mechanism of the College clock, causing it to chime 26 times, or, best of all, dismantled a small car, ingeniously smuggled the parts by lift to the roof of the Government Life building in Cathedral Square and there reassembled it for all the world to see. A report that a bemused traffic officer placed a ticket on it for unauthorised parking[16] only added to the general hilarity. Capping jollification did not come to an end when the Engineering School moved to Ilam, but became terminally enfeebled when Rolleston House closed, though as will be seen, that was not the only cause.

Of the 66 presidents of the Students' Association by 1985, eleven were engineering students. In the last fifteen years however, no engineer has held the position, a change that lends support to the belief that the engineering degree has become a difficult one. But there are other causes too. In 1967, the *Student Engineer*, questioned whether Ensoc had much

*Taylor, James Bruce Craig; b 1915; educ CUC (BE hons), CUC (M.Sc); engg research DSIR, civ & mech engg branches NZR; sen lect UC, 1967-80.

**A greater appreciation of Maori sensibilities led to the demise of the party. The Student Representation Committee in 1971 wanted to disband the haka party, but the Capping Committee regarded the decision as invalid. The 25 grass skirts — actually they were raffia — were thereupon stolen. The party seldom reappeared in subsequent years as Capping Week festivities faltered and died. In Auckland, Maori gang members fell on a similar student haka party in 1979 and assaulted its members.

purpose because the rest of the University was moving to Ilam. At the beginning of the sixties, engineering students felt isolated and looked to Ensoc for assistance that should have come from the Students' Association. Ensoc had, it said, carried out the task satisfactorily, but engineering students were no longer a united body. "This is made particularly evident," it added, "with the disappearance of the 'block vote' for engineers in Students' Association elections."[17] Whatever the cause, engineering students subsequently played minimal roles in student politics, in sharp contrast to 1954, when president, vice-president, secretary and most of the executive were engineering students.

A remarkable change in student attitudes and behaviour occurred between the construction of the Engineering School in 1958 and the opening of the Science buildings in 1966. When Cobham laid the foundation stone for the School, a group of engineering students much enlivened the rather dismal occasion. The Governor-General was escorted by a team of "marching girls", who made a deep obeisance as he reached the stage. Four large "bottles", with a student inside each, formed a guard of honour and a huge papier mache head mounted on the back of a truck nodded sagely at every profound utterance from the stage. When Cobham's successor arrived to open the Science buildings, eight years later he was confronted with a blazing Union Jack (not, let it be said, a gesture by engineering students).

The behaviour of engineering students does not, however, vary much and they are sometimes charged with being the redneck country cousins of the sophisticates of the Arts faculty. The evidence is slight but occasionally their behaviour en masse conforms to the stereotype. When it does, only a small proportion of students is involved; and usually it does not. Observation indicates that engineering students borrow library books on non-engineering topics, are represented in the audience at lunch-time concerts and survive exposure to painting and pottery exhibitions in the Engineering Library. They dress at least as tidily as Law students; one even demonstrated a remarkable grasp of the law in a protracted — and successful — appeal against a Disciplinary Committee finding. Some play games other than rugby.

If they are conservative in outlook, so too is the School and the profession. In 1963, before a wave of unrest swept through the universities of Europe and North America, Ensoc was chided by the consultants' division of the Institution of Engineers. It claimed that part of an employment survey undertaken by a student, John B. Matthews, for a Templin Scroll paper, placed too much emphasis on what it called "socialist ideals". The consultants' claim was described as rash by the *Student Engineer.* Only four of 15 questions, it said, dealt with salary and fringe benefits and in any case "any tendency towards socialistic ideals is quite natural since we have been brought up in a state in which great emphasis is placed on social security."[18] And the Ensoc annual report put it more sharply:

> It was unfortunate that the enlarged questionnaire was misconstrued to provide ammunition for an attack on the School and the Society by members of the profession who should have known better.[19]

A contrary view of students was taken by Dr W.B. Sutch, then Secretary of Industries and Commerce, after engineering students questioned him about the wisdom of establishing industries in New Zealand. The students, he told Canterbury manufacturers later, looked as if they were Neanderthal men who had come out of the caves.

> These young engineering students in the 21-22 years age group (he continued) should be looking forward to finding their fortune in industry, particularly in branches such as industrial designing. But they are going through university with a serious misunderstanding of industry.[20]

The two criticisms indicated that engineering students were keeping solidly to the centre and in any event the *Student Engineer* suggested that Sutch had misunderstood them. The degree course, it said, covered a considerable range of subjects and some aspects of engineering had to be excluded. General information about industry and manufacturing in New Zealand had therefore to be acquired outside lectures.[21]

But there were other indicators of engineering conservatism as the wave of student unrest overseas sent some ripples around New Zealand universities. Student radicalism was expressed by the installation of a contraceptive vending machine in the Student Union, a decision which earned a rebuke from the Prime Minister as well as the University Council, and there was a distinctly more irreverent tone in the student newspaper *Canta.* Students in turn opposed the installation of an Omega navigational system in Canterbury, American (and New Zealand) involvement in Vietnam and an All Black tour of South Africa. The

School was at Ilam by the time most of these events occurred, but it is suspected that few engineers would have joined the protesters in any case.

The University administration was concerned at one of the first manifestations of student protest—opposition to the conferring of an honorary LL.D. degree on a retiring Governor-General, Sir Bernard Fergusson, in 1966. During the ceremony R.H. and College House students, many of them engineers, milled about outside the University Hall, overwhelming any opposition with robust chants of "We love/the Gov" and "We agree with the LL.D."

If engineers took little part in politics and protest, they seemed to be less involved with pot also. A survey in 1971 showed 15 per cent of students used drugs, mostly marijuana (alcohol was not included). Women used pot as much as men. Arts students were more likely to be users than either engineering or science students and those living in flats were more likely to be users than those in private board[22] — and presumably those in halls.

All these factors contributed to a picture of engineers as the conservative sector of Canterbury students; and the picture was given colour by Burton when he told Ensoc in 1969:

> Although you look revolting, you are not revolting enough . . . The University is not the place to be a rebel against society, but a place to learn about being a rebel, a nonconformist. If you come out with only a licence to practise engineering, you have been wasting your time.[23]

That advice was not heeded and familiar complaints of student apathy began to appear in the mid-sixties. Initially though, engineering students were pleased with Ilam, their major regret being the lack of a gymnasium, a facility used by 80 per cent of them on the city site.[24] There was a bonus for Ensoc: for the first time it had an office for the sale of stationery and old examination papers. But if Ensoc retained its members, only 10 per cent turned up to its meetings, which were switched from evening to midday functions in a bid to keep up attendances. Clearly the degree course was taking more and more time to master. Ensoc's 69th annual report complained that "the recent trend of disinterest (sic) among students at the School . . . has become critical." By 1967 the *Student Engineer* was under threat and it was decided to reduce its size by eliminating most of the technical papers it used to publish. In the following year the editor tried to spark controversy by declaring that the degree course needed a wider range of subjects, like philosophy and sociology, to make students aware that engineering is living . . . "the staid, conservative engineer has not, even in his training, been able to get any insight into the relationship of humanity to his creations. Balance . . . has not been achieved."[25] Alas, no one took up the argument and the 1969 issue announced that there was no editor and no business manager for the publication. There was no money either and that edition was the last, though Ensoc plans to publish a special issue in 1987.

But it was not apathy so much as a new study regime affecting all students that reduced extra-curricular activity. The student revue and, mercifully, *Capping Mag* disappeared in the early seventies. Procesh was on again, off again and finally off permanently, though students helped to enliven Cathedral Square subsequently by inviting the Head of the Department of Levity at Melbourne University to visit Christchurch. Ian Brackenbury-Channell liked what he saw when he arrived for the Universities Arts Festival in 1973 and became a more or less permanent feature of Christchurch. Procesh had a final fling in 1978, but it was "monumental for its lack of wit, emasculated by its lack of vitality and was only a shadow of its former self."[26]

Student radicalism was all over by then anyway. "Where have all the angry young men and women gone?" asked the *Press*. The answer was plain. It was Vietnam more than any other issue that caused such radicalisation of the campus as did occur. Student protest had not been ineffective. Now two new phenomena were dominating student attention. One was in-term, year-long assessment, introduced at the request of students, and the other was the spectre of unemployment. Together they stultified traditional student activities.

But Ensoc continues, though at a lower level of activity than in its palmy days, and is more relevant to the immediate needs of its own members than the Students' Association. It silenced critics of its alleged sexist slant too. Joanne Lund, who was elected president of Ensoc for 1983, was the first woman to hold the office, but she regarded it as unremarkable and hoped there would be "no fuss" about her election.[28] A year later, Mrs Gretchen Kivell (nee Kershaw), a 1970 graduate in chemical engineering, was elected chairman of the Auckland branch of IPENZ. These were only the first cracks in the wall of male dominance of engineering; but are not the first cracks the most important?

CHAPTER EIGHT

Retrospect and Prospect

In its first 100 years the School might not have produced a Rutherford, a Condliffe, a Kerr or a Tinsley, but it did produce 6143 graduates, all but 85 of them male. In addition 666 higher degrees (478 M.E. and 188 Ph.D) have been conferred, giving a grand total of 6809 degrees. The hundredth B.E. was not conferred until 1926, nearly 40 years after the School's establishment. The thousandth degree was in 1956 and a total of 1248 was conferred while the School was still on the city site. Since the move to Ilam 26 years ago, 4895 first degrees, nearly 80 per cent of the total, have been conferred.

Until Auckland's graduates began to appear, the Canterbury graduates were largely responsible for the physical development of New Zealand. Roads, railways, bridges, ports and airports, electricity generation and transmission, sewerage, communications and the growth of the cities were monuments to their skill, expertise, imagination and training. Few young graduates were entrepreneurs with promise of meteoric careers in industry and commerce. In contrast to other professions, the School's graduates were more likely to become employees of local or national government, a company or a partnership. Doctors and lawyers had more of a tradition of hanging out their own shingle; moreover the public was — and still is — less sure about engineering than about medicine and the law. Most people have occasion to consult a doctor or lawyer during their lives, but few ever consult an engineer and the public debt to the profession is not widely appreciated.

Lack of appreciation does not however obscure the competence of the Canterbury engineers and their splendid record locally and nationally in peace and in war. A random survey suggests that many made their mark outside New Zealand too. Indeed the first graduate (in mechanical engineering) in 1896 — one of the seven children of Bishop Churchill

Julius - went to Australia and there, among other things, developed that essential appurtenance of Australian and New Zealand life, the mechanical totalisator. George Alfred (later Sir George) Julius, was also head of the Commonwealth Scientific and Industrial Organisation 1926-45. Archibald Milne Hamilton, who worked for the Lyttelton Harbour Board after graduating in 1924, went to the Admiralty, helped design the Singapore naval base and devised the Hamilton bridge, which was adopted by the War Office in the Second World War and used in many parts of the world for the reconstruction of bridges.

Eric Norman Webb (1889-1984) had no sooner gained his associate qualification in civil engineering than he became chief magnetician for Sir Douglas Mawson's Antarctic expedition of 1911-14. The quality of his work was admired by Antarctic scientists nearly 70 years later[1] and contributed to determining the exact location of the South magnetic pole.

Sir Hector McNeil, who graduated B.E. in 1929, joined the British firm of Babcock and Wilcox Ltd, London, after working for the Victoria State Electricity Commission and eventually became managing director as well as a director of eight associated companies overseas. James Collett Dickinson (B.E. 1922) became a consulting engineer specialising in hydro-electric development after being associated with construction of the Sydney Harbour and Newcastle bridges, the Nag Hammadi barrage in Egypt and the Singapore naval base. M.W. Mehaffey (B.E. 1910) was chief engineer for Australia and was responsible for notable engineering works like the Captain Cook graving dock in Sydney. William Wilson (B.E. elect. 1906 and D.Sc. 1936) was chief of development for the General Electric Company, England; George Clifford James Dalton (B.Sc. 1937 and B.E. 1939), a Rhodes scholar, contributed to British nuclear research before returning to a chair of mechanical engineering at Auckland; and Dr W.I. Axford, F.R.S., who graduated M.E. and M.Sc. from Canterbury and Ph.D. from Manchester, became a director of the Max Planck Institute for Aeronomy, Katlenburg-Lindau, and was Vice-Chancellor of Victoria University of Wellington. He is a distinguished astrophysicist who has taken an important part in space exploration.

Many graduates achieved academic eminence. Seven of the 13 chairs in the School in 1986 were held by graduates. R.G.H. Prince, one of the School's most brilliant graduates, was on the Chemical Engineering staff before taking the chair in that subject at the University of

Queensland. A.R. Acheson (B.E. and B.Sc. 1906) became a professor of engineering at Syracuse University. Graham Powell (Ph.D.1965) had a similiar position at the University of California, Berkeley. Many of the senior staff of the Auckland school are also Canterbury graduates. They include the Dean, R.F. Meyer (B.E. 1954); the head of electrical engineering, J.L. Woodward (B.E. 1948), who was acting-head of the department at Canterbury before Kay's appointment; N.A. Mowbray, a former professor of civil engineering (1951-76) who graduated B.E. in 1937 and who also lectured at Canterbury; and A.L. Titchener, who held a chair in mechanical engineering and subsequently in chemical and materials engineering.

Graduates filled numerous key positions in New Zealand. F.R. Askin, (B.E. 1934. and Hon. D.Sc. in 1981) and R.G. Norman (B.E. 1945) were Commissioners of Works. Numerous power stations were designed and built under the direction of F.T.M. Kissell (B.Sc. in mechanical engineering in 1905), who was general manager of the State Hydro-electric Department, and W.M. Duncan (B.E. 1944) became Secretary of Energy. J.P. Hollings (B.E. 1950), senior director of New Zealand's largest consulting firm, is a leader in structural engineering who made a significant contribution to earthquake engineering.

Numerous graduates became engineers for local authorities as well as for the major engineering works that opened in the prosperous post-war decades. Oddly, the School's greatest benefactor was not even one of its graduates. John Angus Erskine (1872-1960), a contemporary of Rutherford, graduated M.A. in 1895 before completing his engineering education in Germany and on his death left more than half a million dollars to the University to enable teachers of engineering, science and commerce to travel overseas and to bring experts in those fields to the University. Erskine's portrait, by W.A. Sutton (he had only two snapshots as a guide and kept the work "very loose") hangs in the School and scores of engineers have cause to remember him with gratitude if not affection.*

The School's location on the campus and the heavy workload of students combined to isolate the engineers from the student body but other factors have prevented any deep division. Students who do not gain direct entry

*Erskine, a bachelor, would probably not have sparked much affection during his lifetime if his austere advice on bringing up the young was any guide to his character: "Every child (he said) should have dominoes as early as possible to teach him (sic) to count and from the age of seven he should play chess to teach him to think."[2] Erskine was twice N.Z. chess champion.

experience a year in the Science Faculty and some extend that association by taking courses in engineering geology and chemical process technology as well as options in other faculties. The first course in engineering geology was established in 1966 and was replaced by an M.Sc. course in 1974. Subsequently a one-year diploma was established for graduates of either engineering or geology to enable them to evaluate geological hazards and problems. Graduate students also learned about other disciplines; and another common interest is computing, which at Canterbury began in the School. In fact the School had scarcely settled at Ilam before it was precipitated into the computer age.

Canterbury was the first New Zealand university to order a computer and the IBM 1620 installed in the Electrical Engineering Department in April 1962 for University-wide use was the first in the South Island. The negotiations, ordering and arrival of the big, high-speed, fully-transistorised, electronic, digital (sic) computer was reported almost breathlessly by the *Press*, which described how it was delivered to its laboratory in Rooms F202 and 204 with less trouble than a grand piano would have caused.*[3] Bruce Moon, applied science representative for IBM, was appointed a senior lecturer in mathematics to preside over the new facility, officially the Mobil Computer Laboratory, and Mrs Joan Lester was secretary and punch card operator.

Llewellyn expected "an avalanche" of demand and he was right. Within three months the computer was in use for 65 hours a week and in five months 1000 visitors had marvelled at it. Extension courses attracted large enrolments and at the end of 1964 a seminar on applications of digital computers in the process industries was packed with practising engineers. Indeed Pownall told the University Council that the University was becoming embarrassed by the load on the computer of between 19.5 and 24 hours a day, even after its memory had been doubled. When the Rev. W.M. Hendrie commented that Ph.D.s had been awarded before computers were even thought of, the Vice-Chancellor said: "We are in the computer age; even our undergraduate students are being trained for it."

"I'm glad" , Mr Hendrie replied, "that I was born when I was."[4]

Few engineers harboured such unworthy thoughts, but the very success

*Purchase of the computer, which cost the equivalent of $54,000, a considerable sum in 1962, was made possible by a substantial discount and contributions from Mobil Oil New Zealand Ltd.

of the laboratory led to its removal from the School. The computer was taken from its first home in September 1966 and recommissioned in the new Computer Centre. Thereafter engineers (along with mathematicians and scientists) trailed across the campus, sometimes in gumboots, with a new symbol of academic status tucked under one arm — a box full of punch cards. The 1620 was replaced in 1967 by an IBM 360/44, said to have 60 times the speed and capacity of its predecessor, and computing became a substantial University-wide activity. But only about 8 per cent of undergraduates were involved at that time; computing was a research tool. Engineering, science and commerce staff and research students were major users; but by 1977, 26 of 30 academic departments were making use of computers, which took some 3 per cent of the University's total budget. The 360/44 was replaced by a Burroughs B6700 and that yielded in turn to a Prime 750 and a B6930, installation of which enabled an interactive network to be developed on the campus. After that came the deluge of micro-computers. The gap between the present and the future seemed to shrink ever more rapidly.

When A.H. Barth built an analogue computer on the city site in 1957, it made little impact and proved to be inadequate for subsequent needs. But MacElwee, whose principal interest was in electrical power and machines, saw the need for an analogue computer, and the first to be purchased, an FR48, was installed in November 1963. Electrical Engineering was also the first department to purchase its own computer, an EAI640 installed in 1968. It was interfaced a year later with an EAI580 analogue computer and this was the only hybrid computer in New Zealand. It too was virtually a research tool and was used intensively for modelling engineering problems.

In 1970 Civil Engineering joined forces with the School of Forestry to buy a PDP-8 and since then each engineering department has bought computers to increase computing power within departments to levels not imagined for the whole University a few years before. Access to the new mainframe computers in the Computer Centre was provided by a network exchange developed in the eighties and nearly 40 terminals were provided in the School, giving users access 24 hours a day if required. The School's mini-computers delivered as much "crunch" as the centre could provide. Micro-computers proliferated in every department and computer work stations became as commonplace as pocket calculators had been a few years before. Their use far outstripped the growth in student numbers and stimulated research as never before. Fortunately, technological changes substantially reduced the cost of computers while increasing their capabilities, but costs remained a recurring problem which put a brake on many desirable developments.

By 1986, with the future of computing far from clear, the wheel had turned almost full circle to a single user system. There were, however, two remarkable differences. First, enormous computing power could be delivered to individual users, enabling them to undertake interactive computing, especially in graphics, that could not previously have been contemplated. Users whose requirements exceeded the campus capacity could make use of large American systems at low cost by satellite. The second major difference was the rapid increase in the inclusion of computing in undergraduate teaching. Computer literacy certainly distinguished the class of '85 from that of '75. Later students were at ease with computers, knew how and when to use them and when (perhaps as importantly) not to use them. It was widely acknowledged that the universities, especially the engineering schools, required substantial resources for computer re-equipment in 1986 and Canterbury's engineers hope they might arrive as a bonus to the centennial celebrations.

In Britain a degree in engineering might be, as the Duke of Edinburgh claimed, less respectable than a degree in political science,[5] but in New Zealand engineers are unquestionably in the professional class, if not at the top of it, and their sons, like those of doctors, often follow them into the profession. Major engineering works are visible, there is an appreciation of the importance of engineering in the national economy, and there is also some recognition of the absolute nature of the discipline: buildings, bridges or aircraft either stay up or fall down whereas the law, business, economics and even medicine are usually more forgiving of small lapses in professional competence.

A further reason for public confidence is the close relationship between School and profession, prompted in part by the decision to make engineering a graduate profession. In 1940, only 17 per cent of 916 registered engineers in New Zealand were graduates and 41 per cent had no formal qualification. By 1982, 77 per cent of 4279 engineers were graduates and the remainder had other qualifications. The latter were principally the result of the examinations conducted by British professional institutions, which were once considered to be of equivalent standard to the degree course. But from the 1960s the standard of the B.E. rose, and though institution examinations were reorganised, IPENZ in 1982 finally withdrew recognition of them. Assuming responsibility for standards has meant close collaboration between IPENZ and the School and regular moderation now takes place to ensure a balance between academic offerings and professional and industrial requirements.

There are in fact few differences. Academic and professional engineers recognise that while the future course of engineering, especially electronics, is difficult to predict, it will be beyond the grasp of those lacking a sound training in fundamental principles. The aim, according to the institution's education policy, is to give that understanding through broad and deep courses in basic scientific subjects, mathematics and the principles of engineering. In the words of the policy statement:

> The advances of engineering are related to the advances of the sciences. Engineers must be able to act as applied scientists. To be able to co-operate on equal terms with the pure scientists they must have a comparable intellectual status and a shared education background.[6]

The School and profession collaborate in other ways. Practising engineers sit on the Faculty and numerous academic engineers have held office in the institution. The institution assists with the organisation of public days and open days for schools and provides funds for brochures and a film on engineering education. The profession also lent support to the School in its drives for buildings, equipment and the appointment of technicians, who have served the School well. There is an easy familiarity betwen academic, library, technical and administrative staff that has given the whole enterprise a relaxed yet purposeful atmosphere in which all staff members can contribute their knowledge, expertise and skills. George Clark, a former senior technical officer, exemplified something of technician enterprise while working with Kay on the design of transducers for the blind aid. He saw the need for a collapsible long cane for use by the blind and devised an effective jointing mechanism which overcame the disadvantages of previous designs.

But it was in refresher work that School and profession really met. If the B.E. degree was the major requirement of a licence to practise, there was recognition of the need to renew that licence as engineering became more complex. The speed of technological advance revolutionised engineering practices in the second half of the century and the Institution investigated the possibility of formalising continuing education for its members. In the end, it left the task largely to the two engineering schools; and it was certainly not disappointed in the Canterbury contribution. As early as 1964 the consultants' division of the Institution complimented the University on its promotion of closer relations between scientists and technologists on one hand and the public on the other, and in particular on its refresher courses.[7]

It was Hopkins who, as in most other aspects of engineering education, led the way by inviting public works design and construction engineers to a week-long seminar on concrete and soils in 1953. A former Works Commissioner recalled:

> It was his (Hopkins's) early understanding of the value of continuing education which led to the remarkable relationship we now enjoy between the practice of engineering and its institutional teaching process. From those humble beginnings have sprung almost two generations of practitioners for whom the campus is not a finite terminal element at the beginning, but remains an ever-present environment with its own dynamics and interactions with the workplace. That was the legacy of Harry Hopkins and those who shared in it will always be his debtors.[8]

In 1962, Frank Lu* was awarded the W.A. Stevenson prize of the NZ Institution of Engineers and went to the United States to investigate the critical path method of construction planning developed by the U.S. Navy to accelerate its Polaris missile programme. He was an immediate convert and returned to New Zealand with an almost missionary zeal to initiate critical path planning in the construction industry in particular and in industrial and business management generally. He presented papers to Institution branches, gave numerous interviews and even wrote a book on the topic — *The Critical Path Method of Construction Management* (Christchurch, 1964). But it was the seminars he conducted for practising engineers and subsequent refresher courses that ensured swift adoption of the method in New Zealand. They were attended by hundreds of engineers, contractors and businessmen and demonstrated how swiftly and effectively new ideas, methods and processes could be widely disseminated. Otago University was so impressed that it offered Lu a chair in management in 1972, though he had no formal qualifications in economics, accountancy, management or administration.

---

*Lu, Frank Pao-shan, 1925-84; b China; educ St John's U, Shanghai (B.Sc. in civ engg); bridge design NZMOW 1951-57, s.lect civ engg CU 1958, prof. management Otago U. 1970-72, Sir James Fletcher chair ind.admin. CU 1972-84.

Similar seminars, organised by the Extension Studies (subsequently Continuing Education) Department introduced numerous engineering innovations to the profession. A series of two and three-day seminars on microcomputer fundamentals and applications in the late seventies, for example, was attended by 400 engineers and technical managers and had a major impact on the speed of adoption of microcomputers within New Zealand.[9] The campus was increasingly enlivened during University vacations by well-dressed and labelled practising engineers learning about applied digital image processing, microprocessor controls in process industries, harmonics in power systems, concrete construction design or VLSI circuit design. And it was not only professional engineers who benefited. An all-male class of 140 radio technicians and enthusiasts crowded six evening classes by Withers in 1962 on *Techniques of Transistors* and not one of them missed a lecture.[10]

It was demanding work and the Engineering faculty established a sub-committee to investigate the extent to which staff should be involved. It acknowledged the School's responsibility - though not sole responsibility - for the further education and re-education of professional engineers. The Faculty subsequently welcomed recognition by the Academic Policy Committee of the responsibilities of the professional schools and Extension Studies Department in the provision of post-graduation education.[11]

The profession's response to extension work at that level was welcome because the esoteric nature of some research, especially in electronics, has posed the question of whether it is becoming too remote from an engineering context. Indeed a former City Engineer, Peter Scoular, complained some 20 years ago that engineering courses were becoming so scientific that the B.E. could well be called a degree in engineering science.[12] He and other engineers drew a clear distinction between science and their profession: "The central business of science is research, whose product is knowledge. The central business of engineering is design . . . Design encompasses development and often invention, but differs totally from research."[13]

Scientists, it has been said, explore while engineers create. Scientists solve the problems they can, while engineers solve the problems they have to. But the distinction is between the "pure" scientist and the engineer. Few would disagree with the concept of an engineer as an applied scientist. The applications of science in the School have, as has been seen, been significant; and continued research at that level is more likely to advance engineering than limit it. One organisation, the Royal Society of New Zealand, recognised the scientific input of the School's leading researchers and elected six of them to fellowship: Kay (1971), Bates (1976), Park (1978), McCallion (1980), Paulay (1983) and Arrillaga (1986). Only one other New Zealand engineer, also a Canterbury graduate, has been so honoured.

In what ways will the School develop as its second century begins? It would be rash to look far ahead for in science and technology the only certainty is that there will be substantial and rapid change, especially in electronics. Although restrictions on entry to the School were approved by the University Council a quarter of a century ago in 1962, they have been imposed principally to regulate numbers of overseas students and to control entry to the Electrical and Electronics Engineering Department. The department's expertise is widely acknowledged and with additional staff it could provide useful service courses for science departments, such as Physics and Computer Science; and some even predict that a marriage with the latter department might eventually be arranged. Extension of laboratory and office space cannot long be delayed; electronics is likely to develop at even greater speed than in the past.

Civil Engineering might have become the No. 2 department in size, but it will continue to grow; and because of a greater number of elective subjects, especially in the third professional year, laboratory space in particular, will need to be expanded. Research on special New Zealand projects, for which solutions are sought by both industry and Government departments, also demands additional laboratory space. That research will not only be in traditional civil engineering fields. One new development, known as "retrofitting", enables old buildings to be preserved rather than demolished. The process involves pouring concrete walls, which meet current building safety codes, inside buildings which do not, and thus extending the life of the buildings indefinitely. The process was devised not only for reasons of economy: it also enables useful, handsome or historic buildings to be preserved and used.

An inevitable consequence of new courses in the Chemical and Process Engineering Department will be additional laboratory space. The New Zealand interest in self-sufficiency in liquid fuels in future will stimulate the study of fuel technology, while public concern with the safety of large process facilities handling potentially dangerous materials will require extension of courses in loss-prevention technology and reliability in the process industries. A laboratory for teaching and research in hazardous operations will be an essential development. In addition, some chemical engineering graduates (with others in chemistry and forestry) will be specially fitted to bridge the gap between the development techniques of molecular biology and the application of those techniques.

The Mechanical Engineering Department was the first to impose a restriction on entry into the first professional year because of cramped laboratory conditions; and though the roll has steadied, the laboratories remain crowded. Additional space, in particular, is required for work in production technology. The department also requires a share of a hazardous operations laboratory for its foundry and heat treatment work and extension of its instrumentation and control work is inevitable. Some predictions are that the need for close co-operation between electrical and mechanical engineers will result in a joint degree course.

The U.G.C. in 1962 recommended that the School should be extended to 200,000 square feet for 700 students (it actually got 205,200 square feet (19,067 square metres). The only subsequent extension was (as has been seen) in the Library. But since 1968, the roll has grown far above 700 and there has been consequent pressure on laboratory, lecture theatre and drawing office space. The higher roll also resulted in the appointment of additional staff, who, while stimulating research, also put additional pressure on laboratories and office space. However the School develops in future, expansion is inevitable. The School has sought extensions of about 40 per cent of current space — an additional 7780 square metres. If it expands to cater for 1200 students, it will have an annual intake of about 320, with about 100 going into the electrical and electronic field.

But new courses, equipment and buildings and growth of the student roll are not the only items on the agenda for the second century. The School is also looking outwards. To mark the centennial, industry and profession have been asked to contribute $2.5 million over the next five years for the establishment of a Centre for Advanced Engineering at Ilam. The centre, on the lines of the Warren Centre at the University of Sydney, will bring project groups of engineers from industry and universities together to work for a brief, intensive period on an engineering problem of national importance. It will not be a research centre but a centre for the transfer of technology to foster excellence in engineering for the benefit of industry, academic engineers and the whole country.

The University was also considering a proposal from the Engineering Faculty towards the end of 1986 for the establishment of an office, tentatively called Canterprise Ltd., to service inquiries for staff research, testing and investigations, consulting by staff and the use of University facilities as well as to commercialise high technology innovations and inventions arising from research in the University. Canterbury has been slow to move on such a proposal, but the "user pays" climate emphasises the advantages of putting departments, especially those in the School, on a more commercial footing.

These proposals suggest that the School may become even more dynamic and innovative at the start of its second century and will not look back with much nostalgia to the cloisters and quadrangles of the city site. In any case, for a majority of graduates attending the centennial celebrations and for most staff, the School's home is Ilam. But there is a substantial difference between the School of the early sixties and that of 1986. The Library, the research activity, computers, terminal clusters, equipment and the calibre and number of staff suggest that the School is not only shaping up to a new century, but is preparing to shape it as well. That was not the case when the first academic pilgrims came to Ilam. Then the acting-Librarian, K.A.R. Horn, reported that the spare key to the Ilam homestead was missing from where it was kept —

> namely on the trunk of the tree near the front door . . .
> I suggest a better hiding place because though the present place is thought to be something of a secret, it is in fact known to a large number of people who have no real authority to know about it.[14]

And the bursar complained about mail for the School lying about the grounds because "apparently postmen place the mail in a milkbox which has no back and is blown out".[15] Better arrangements for keys and mail, as well as for most other facets of School and University life, have been devised subsequently. Indeed 1960 now seems hardly less distant than the heyday of Scott. Academics might still complain about some details, but the developments of the last 25 years have been as swift as they have been far-reaching. They have created a sense of purpose and high staff morale. Could the School ask for much more as it enters its second century?

Appendices

APPENDIX 1

Engineering School Staff

Academic Staff Pre-1929

Cook, C.H.H. 1887-9 supervisor
Scott, R.J. 1888-23, i/c eng 1890,
  p i/c Eng & Tech Sci 1894,
  p i/c El Eng 1900-10
Dobson, E. 1888-92 l civ eng 1888
Morrison, A.F. 1893-1928, l 1923
Craddock, A.R. 1901-2 l 1901
Farr, C.C. 1904-10 l 1904
Powell, P.H. 1905-29 p elec eng 1923
Steele, S. 1905-7, 1908-28 p mech eng 1923
Robinson, W.F. 1910-28 l 1910
Bamford, P.G. 1916-28 res asst 1921
Campbell, R.A. 1920-8 p civ eng 1923
Calvert, G.G. 1921-8 l 1926
Pollard, T.R. 1922-8 l 1926
Pennington, J.H. 1923-5 l 1924
Grundy, R.H. 1924-8 l 1926

Academic Staff By Departments After 1929

Chemical and Process Engineering

Siemon, S.R. 1952-65 sl & hd 1953 p & hd
  1957
Hagyard, T. 1952-71 r 1963
Prince, R.G.H. 1958-60 sl 1960
Smith, S.J.C. 1960-63 l 1960
Peet, N.J. 1961- sl 1971
Earl, W.B. 1962- r 1986
Keey, R.B. 1962- p 1978; hd 1981-85
Kennedy, A.M. 1965- p & hd 1965-81
Stott, J.B. 1965- r 1974
Allen, R.M. 1966- sl 1973
Williamson, A.G. 1967- p 1974 hd 1985-
Abrahamson, J. 1970- sl 1976
Gilmour, I.A. 1970- sl 1975
McElroy, P.J. 1971- sl 1974
Graham, E.E. 1972-1981 sl 1976
Jordan, P.J. 1972- sl 1985
Kerdemelidis, C. 1981-83 1 1981

Civil Engineering

Calvert, G.G. 1929-50 p & hd 1946
Robinson, W.F. 1929-35 1 1929
Cull, J.E.L. 1930-42 p & hd 1930
Stanton, T.M. 1940-5 sl 1945
Alley, P.J. 1946-66 r 1966
Massey, P.C. 1947-55 sl 1952
Owen, B.T.W. 1947-53 l 1947
Dunn, P.H. 1948 sl 1948
Mowbray, N.A. 1949-51 sl 1949
Gilmour, P.M. 1949-50 l 1949
Smith, D.B. 1950-1 sl 1950
Fraser, P.E. 1951-2 l 1951
Giddens, E.P. 1951- sl 1963
Hopkins, H.J. 1951-78 p & hd 1951
Belshaw, T. 1952-3 l 1952
Henderson, F.M. 1952-68 p 1964
Holmes, I.L. 1952-7 sl 1952
Simpson, A.H. 1952-66 sl 1952
Williman, A. 1952-80 sl 1953
Neville, A.M. 1954-5 l 1954

Scrivener, J.C. 1955-77 r 1970
Park, R. 1956-9; 1965- p 1968
Lu, F.P.S. 1958-70 r 1969
Arnold, M. 1959-60 l 1959
Shepherd, R. 1959-71 r 1968
Yao K.M. 1960-9 sl 1966
Dodd, T.A.H. 1961- sl 1967
Paulay, T. 1961- p 1975
Powell, G.H. 1961-5 l 1961
Glanville, J.I. 1962-5 l 1962
Smith, N.A.F. 1962-5 l 1962
Elms, D.G. 1964- p 1978
Gordon, D.R. 1967- sl 1973
Hince, R.F. 1967-71 sl 1967
Moss, P.J. 1967- sl 1970
Sutherland, A.J. 1967- r 1973
Carr, A.J. 1968- sl 1972
Blakeley, J.P. 1969-70 l 1969
Fellows, A.G. 1971-77 sl 1971
Gaerty, L. 1971-82 sl 1979
Wood, I.R. 1971- p 1971
Cooke, N. 1972- sl 1972
Davis, R.O. 1972- r 1979
Hunt, B.W. 1972- r 1978
Walpole, W.R. 1973- sl 1980
Priestley, M.J.N. 1975-87, r 1978
Seddon, P.A. 1975-84, sl 1975
Wilson, F. 1976- sl 1976
Berrill, J.B. 1977- r 1986
Dean, J.A. 1978- sl 1981
Scott, D. 1978- sl 1978
Spigel, R.H. 1978- sl 1983
Mullenger, G. 1979- sl 1979
Nicholson, A.J. 1981- sl 1981
Pidwerbesky, B.D. 1986- l 1986

Electrical Engineering

Powell, P.H. 1929-47 p & hd 1929
Pollard, T.R. 1929-48 sl 1941
Neale, R.D. 1940-62 sl 1947
Wright, W.S. 1937-52 l 1947
Withers, B.T. 1946-71 sl 1950
MacElwee, N.M. 1947-65 p & hd 1947
Bowen, W.H. 1948-75 sl 1948
Kennedy, W.K. 1963- sl 1975
Byers, D.J. 1964- sl 1970
Watson, P.W. 1964-75 sl 1972
Jelinek, H.J. 1965-9 l 1965
Kay, L. 1965-86 p & hd 1965-82
Andreae, J.H. 1966- p 1966 r 1973
Johnson, D.A.H. 1966-68 l 1966
Bates, R.H.T. 1967- r 1968 p 1975
Castle, N.J. 1967-8 l 1967
Calhaem, N.L. 1951-4 sl 1954
Stephen, K. 1951-64 sl 1960
Woodward, J.L. 1953-65 r 1964
Barth, A.H. 1955-82 sl 1965
Currie, C.V. 1962-63 l 1962
Bargh, J.K. 1963- r 1971 p 1973 hd 1982-85
Gibson, J.A. 1967- r 1974
Dew, J.G. 1968-70 l 1968
Kerdemelidis, V. 1968- sl 1968
Grimshaw, K.P. 1969-85 sl 1969
Edward, L.N.M. 1970- sl 1973
Cady, F.M. 1971-80, sl 1979
Haines, K.A. 1971-73 r 1971
Sirisena, H.R. 1971- r 1980
Undrill, J.M. 1971-2 sl 1971
Watson, D.B. 1973, sl 1973
Webb, J.A. 1974- sl 1974
Squires, P.L. 1973- l 1973
Boys, J.T. 1974-77 sl 1977
Errington, J.G. 1978-80 sl 1978
Tichener, M.R. 1982-83 sl 1982
Hodgson, R.M. 1975- sl 1979
Arrillaga, J. 1975- p 75, hd 1985
Arnold, C. 1976- sl 1981
Gough, P.T. 1980- sl 1983
Bodger, P.S. 1982- sl 1985
Dewe, M.B. 1978-81; 1985- sl 1985
Duke, R.M. 1984- l
Garden, K.L. 1985- l

Mechanical Engineering

Steele, S. 1929-46 p & hd 1929
Bamford, P.G. 1929-46 l 1941
Grundy, R.H. 1929-32 l 1929
Morrison, A.F. 1929-32 l 1929
Rastrick, R.J. 1929-68 p & hd 1946
Lush, A. 1936-54 sl 1950
Hewitt, N.E. 1949-69 sl 1945
Allfrey, M.A.A. 1945-9 sl 1949
Simpson, A.H. 1946-8 l 1946
Adams, H.F.T. 1948-71 ap 1955
Martin, J.K. 1949-52 l 1949
Mackay, N.A. 1950-2 sl 1950
Newstead, W.F. 1950-73 sl 1971
Burton, A. 1951-71 sl 1951

Rodger, M. 1951-73 sl 1951
Whittle, G. 1951-80 sl 1960
Rainbird, W.J. 1952-4 l 1952
Stevenson, D.C. 1954-63 sl 1954 1969- p 1969 hd 1969-83
Fahy, F.W. 1957- sl 1964
Satterthwaite, C.A.S. 1959-85 sl 1969
Bewley, N.J. 1960-6 l 1960
Dunmore, J.H. 1964 l 1964
Erasmus, L.A. 1965- sl 1968 r 1973 p 1979 hd 1983
Zeff, C. 1965-7 l 1965
Papesch, A.J.G. 1966- l 1966
Taylor, R.M. 1966-80 sl 1970
Taylor, J.B.C. 1968-79 sl 1968
Smaill, J.S. 1969- l 1972 sl 1979
Bowen, A.J. 1970- sl 1976
Lindley, D. 1970-8 sl 1970
McCallion, H. 1971- p 1971
Rhoades, B.L. 1970-9 l 1974
Parker, G.J. 1971- sl 1971
Raine, J.K. 1971-5 al 1971; l 1974; 1977-80 sl 1980; 1984-
Naguleswaran, S. 1972- sl 1975
Whybrew, K. 1972- sl 1978
Astley, R.J. 1973- sl 1978
Burns, N.D. 1973-77 sl 1977
Harman, R.T.C. 1973- l 1974 sl 1977
Tucker, A.S. 1974- sl 1979
Eversman, W. 1975-8 p 1975
Musolff, C.F. 1975-8 l 1975
Britton, G.A. 1979- sl 1985
Dunlop, G.R. 1980- sl 1980
Cowling, J.M. 1981- l 1981
Ditcher, A.K. 1981- l 1981
Green, R.K. 1981- sl 1981
Ilanko, S. 1986- l 1986

Agricultural Engineering

(Based at Lincoln College. The following staff have been associated with B.E. (Agric.) or B.E. (Mech.) teaching since 1963, when Agricultural Technology was first taught as a Mechanical option. B.E.(Agric.) teaching began in 1967.)

Boughton, W.C. 1967-68 sl 1968
Burton, J.R. 1965-70 p 1965
Calbert, I. 1968-87 r
Cherry, N.J. 1975- sl
Chilcott, R.E. 1973- sl
Dakers, A.J. 1979- sl
Davies, T.R. 1975- r
Domogan, I.R. 1986- l
Douglass, B. 1968- sl
Huber, D.G. 1974-82 r
Judd, C.B. 1969-73
Lindsay, G.G. 1963- sl
Mackenzie, D.W. 1970- sl
MacMillan, R.H. 1968-70 sl
Painter, D.J. 1982- r
Shumway, C.A. 1970-73 sl
Van't Woudt, B.D. 1973-78 r
Ward, G.T. 1971- p
Webb, M.E. 1975-77 l

Engineering Library

Engineering Librarians

Jobberns, H. Miss 1946-48
Burke, S. Miss 1948
Jenkins, D.L. 1949-51
Hale, B. Miss 1952-60
Macmillan, F. Miss (Acting) 1961
Shearer, J.E.C. 1962-79
McCarrigan, H. Miss 1979-

Assistants

Dorrell, R. Mrs 1961-2
McEwan, D. Mrs 1961-63
Eskett, J. Miss 1962-70
Tobin, P. Miss 1964-66
Pearson, M. Mrs 1966-69
Blundell, J. Mrs 1966-68
Studholme, J. Miss 1968-69
Barnett, P. Mrs 1969-72
Holt, M. Miss 1979-72
Eng, A.L. 1971-73
McArtney, T. Mrs 1972-73
Guest, M. Miss 1972-73
Jenkins, M. Mrs 1973-74
Lynch, K. Miss 1973-75
McCarrigan, H. Miss 1974-79
Oliver, O. Mrs, 1974
Porter, A. Mrs 1974-75
Harkness, H. Mrs 1974-
Sherley, J. Mrs 1975-76
Savage, D. Miss 1975-

Dickey, P. Miss 1975-77
Chalmers, J. Mrs 1975-76
Conly, F. Mrs 1976-81
Mock, P. Mrs 1976-82
Cummins, S. Miss 1977-79
Thomas, P. Mrs 1977-85
Skidmore, D. Mrs 1979-
Redmayne, J.W. 1980-86
Barrow, J. Mrs 1981-
Gifford, V. Mrs 1981-
Garland, M. Mrs 1983-
Jordan, P. Miss 1986-

Technical Staff

Civil Engineering

Pascoe, W.E. 1950-60
Gibson, O.H. 1951-55
Hudson, R.G. 1952-53
Prebble, N.W. 1953-
Archer, G.F. 1956-81
Watson, H.T. 1956-78
Byers, J.N. 1961-74
Hirst, A.R. 1961-77
Banks, A.R. 1961-67
Hollard, V.D. Miss 1962-63
Sim, G.W. 1963-
Tapper, G.L. 1963
Dawson, P.C. 1964-73
Marrion, K.L. 1964-77
Matheson, R. 1965-66
Gavars, V. 1965-82
Babbonick, F. 1966-68
Foot, A.G. 1966-77
Pearce, H.A. 1966-
Van Dyk, J.G.C. 1967-
Sheard, J.A. 1967-80
Hall, R.P. 1967-
Robinson, P.J. 1968-85
McClelland, W. 1968-84
Crowther, H.H. 1968-
Bennet, C.H. 1969-70
Patterson, H. 1970-80
Robinson, S.R. 1970-73
Adams, J.M. 1971-76
Ballantine, G.P. 1972-74
Hill, G.E. 1973-
Creswell, G.L. 1974
Payne, R.A. 1974-75
Stokes, A.K. 1974-86
Hickey, N.J. 1975-
Mulvey, K.S. 1975-
Robinson, A. 1976-78
MacDougall, N.W. 1976-
Clarke, G.H. 1977-
Allen, R.N. 1977-
Grey, V.J. Mrs 1977-
Bell, A.M. 1978-
Ritchie, J.A.R. 1978-84
Mitchell, P.G. 1979-85
MacPherson, D.J. 1979-
Gardner, L.H. 1980-
Sheppard, I.V. 1981-
Coursey, P.F. 1982-
Scott, T.V. 1984-85
Greenslade, F.R. 1985-
Watt, M.M. 1985
Hutchison, B. 1986-
Dixon, N.A. 1986-
Pasa, S.L.K. 1986-
Schouten, N.A. 1972-74*
Jelly, P.C. 1974-75*
Hobbs, H.J. 1977*
Claridge, J.P. 1978*
Richards, W. 1983*
Somerville, G. 1983*
Karst, T.P. 1983*
Yellowlees, P.M. 1985-86*
Scott, V.A. 1985-*
Simpson, B.M. 1986*

*Temporary

Mechanical Engineering

Roper, F. 1949-62
Retallick, E.D. 1953-83
Stewart, R.I. 1956-81
Johnson, I. 1962-71
Taylor, A.E. 1962-81
Fletcher, A. 1964-66
Sinclair, K. 1964-76
Anink, H. 1966-
Bolt, O. 1966-
Bright, H.J. 1967-86
Somerville, D. 1967-79
Barclay, B. 1969-71
Webb, M. 1969-87
Grey, V. 1970-71

Towell, P.F. 1970-79
Grant, J. McA. 1971-82
Johnson, G.R. 1971-
Mantell, B. 1971-73
Shelton, J.M. 1971-
Gardiner, R. 1972-73
Taylor, M. 1972-75
Simpson, B. 1972
Stott, L.J. 1973-75
Van Kuppevelt, G. 1973-74
Wyatt, T. 1973-74
Rutherford, A. 1973-74
Tinker, R.E. 1974-
Bullock, S. 1975-76
Cheeseman, L.M. 1975-
Hodge, J.G. 1975-77
Fowlds, G.M. 1975-77
Rea, R.D. 1976
Amies, C.S. 1977-
Bolland, N. 1977-
Elder, J. 1977

Flaws, M.J. 1977-
McDonald, R.R. 1977-81
Hindin, P.J. 1979-
Knudsen, C.D. 1979-82
Holibar, M.G. 1980-81
Breurkes, C. 1981-85
Carter, S.A.L. 1981-84
Harris, E.W. 1982-
O'Hagan, P.D. 1981-
McCune, P. 1981
Aicken, S.C. 1983-85
Healy, D.K. 1983-
Oates, G.R. 1983-84
Stewart, N. 1982-85
Claasen, A.B. 1985
McCracken, N.A. 1985-
Stobbs, K.P. 1985-86
Cox, E.W. 1986-
Callaghan, B. 1986
Lucina, M.J. 1986-

Chemical and Process Engineering

Heath, L.T. 1951-84
Grey, G.R. 1977-86
Webb, T.E. 1972-77
Burt, L.J. Miss 1976-78
Rhodes, I.R. 1970-75
Carrell, J.A. 1967-76
Torrens, I.H. 1965-85
Ferguson, D.M. 1977-79
Canterbury, L.A. Miss 1978-80
Hooper, R.D. 1977
Calvert, C.M. 1966-76
Feary, D. 1970-71
Smith, M. 1970-72
Campbell, C.R. 1965-81
Foot, N.W. 1970-
Boyce, C.R. 1981-

Murray, I.J. 1977-
Brown, D.P. 1969-
Berry, T.R. 1972-
Gordon, R.G. 1972-
Earl, W.M. 1979-
Mahon, M.P. 1984-
Allison, H.C.F. 1986-
Wilson, G.B.M. 1985-
Parry, R.O. 1962-65
Stewardson, A. 1963-64
Ritchie, R.A. 1962-67
Johnson, G.R. 1965-69
Gresham, W.K. 1968-69
Church, V.G. 1969-71
Tate, R.W. 1969-70

Electrical and Electronic

Anderson, R.A. 1974-81
Armstrong, P.S. 1976-78
Atkinson, A.J. 1967-78
Atkinson, S.A. 1980-82
Bartkus, S.K. 1979-80
Brittenden, C.E.R. 1961-78
Benger, B.J. 1970-71*
Carey-Smith, C. 1978-80
Buttle, L. 1980-86
Botting, R.J.C. 1981-84
Clark, G.R.S. 1956-79

Davis, T.K. 1971-72
Eichbaum, F.C. 1980-81
Freykberg, S. Ms 1975-78
Haverkort, D. 1977-79
Hydes, W. 1978
Lauder, C.D. 1979-81
Edwards, D. 1983
Hurley, P.B. 1976-83
Hosted, M.B. 1979-85
La Hood, M.S. 1979-83
Festing, D.A. 1976-77

Murray, D.J. 1983-85
Perry, B.A. 1965-67
Rowe, C.H. 1974-78
Shurety, M.R. 1969-76
Stickley, J.A. 1981-85
Tombling, F. 1981-85
Truesdale, J.T. 1972-74
Tyessling, R. 1961-63
Wright, L.C. 1967-70
Cockburn, K. 1946-74
Moir, J.E. 1967-80
Baran, D.R. 1985-
Berry, R.D.E. 1985-
Chadwick, R.C. 1967-
Cox, A.R. 1979-
Cusdin, M.J. 1970-
Earl, W.A. 1984-
Gordon, M. 1964-
Gray, N. 1968-
Hill, L.W. 1964-
Jongens, J. 1968
Mencel, W.Z. 1986-
Robertson, D. Ms 1986-
Sallis, D.J. 1985-
Vernon, A. 1966-
Witbrock, A.S. Ms 1984-
Young, R.H. 1966-
Buttle, L. 1980-86
Sinclair, D.A. 1983-86

*Temporary

Bursars

Unwin, A. Mrs (nee Comerford) 1949-63
Thomson, D.M. 1963-71
Puddle, J.G. 1971-

Bursar's Office Staff

Smith, Pamela 1948*
Vercoe, Dorothy 1949
Priestly, Mrs Jessie 1949-53
Neale, Dorothy 1949-56
Lummis, Edith 1951-55
Cuddon, Barbara 1952-53
Berrill, Lorna 1952-53
Church, Audrey 1953-55
McJannett, Alice 1953-57
Williams, Deidre 1955*
Voot, Mrs D. 1961*
Kelly, Mrs C. 1961-
Olsen, Mrs N. 1961*
Rogers, Mrs June 1961*
Farr, Mrs J. 1963*
Ballantyne, Miss H. 1964*
Gamble, Miss N. 1964*
Vintiner, Mrs I. 1968-69
Scrivener, Mrs K. 1969*
Bain, Miss P. 1963-67
Brown, Miss C. 1969-70
Robertson, Miss N. 1969*
Berry, Mrs V. 1970-71
Noonan, Miss P. 1968-70
McDowell, Mrs I. 1972*
Smith, Miss G.L. 1972*
Ruffell, Miss S.L. 1970-72
Walker, Miss J. 1972*
Jackson, Mrs G.L. 1969-73
Firkin, Mrs D. 1973-74
Spencer, Miss J. 1972-74
Peters, Mrs J.H. 1973-74
McWhinnie, Miss J.M. 1974-75
Thompson, Mrs A.E. 1973-75
Morton, Mrs N. 1975-
McDonald, Miss R. 1976
Jeal, Mrs D. 1978-
Wright, Miss J.C. 1979*
Jensen, Miss E.B. 1975-80
Kelly, Mrs P. 1979-
Beales, Mrs D.H. 1980-

*Retirement date unknown.

Departmental Secretaries

Cotter, Judith, Civil, 1955*
Cooper, Miss M., Mechanical, 1957*
Ball, Mrs D., Civil, 1958-84
Barnard, E., C & P, 1959*
Zenk, Helga, E & E, 1959*
Niven, R., C & P, 1960-77
Keating, Mrs E.M., E & E, 1961-
Perfect, Mrs A.T., Civil, 1964-68
Newman, Mrs A.H., Mechanical, 1964-70
Maginness, Mrs H.M., E & E, 1967-70
Johnstone, Miss M.E., E & E, 1960-73

*Retirement date unknown.

Levicky, Mrs P.C., Mechanical, 1971-72, E & E 1974-83
Watt, Mrs A., Civil, 1972-
Brown, Ms Jean, 1970-72
Ruffell, Miss S.L., Civil, 1972-73
Stout, Mrs Beryl, Mechanical, 1972-76
McCallum, Miss S.F., E & E, 1972-74
Firkin, Mrs D.J., Civil, 1973-74
Field, Miss P.D., Mechanical, 1975-
Savage, Miss A.F., C & P, 1975
Percival, Mrs J.E., Mechanical, 1975-
Dowell, Mrs P.M., Mechanical, 1975-
Carr, Miss A., C & P, 1975-77
Maxwell, Miss S.P., C & P, 1975-77
Kingsbury, Miss J.M., E & E, 1976-77
Haughan, Mrs A., E & E, 1977-
McEntee, Mrs C.M., C & P, 1977-82
Ritchie, Mrs J.E., C & P, 1977-78
Webster, Mrs N.A., Civil, 1977-78
Stewart, Mrs J., Civil, 1978-85
Singh, Mrs B., C & P, 1979-81
Topliss, Miss J., C & P, 1982
Bennett, Miss S.L., C & P, 1982-84
Anderson, Miss L.K., C & P, 1983-
Hartin, Mrs B.A., E & E, 1983-84
Reid, Miss C.M., Civil, 1984-
Daly, Mrs P.L., E & E, 1985-
MacKenzie, Mrs J.M., Civil, 1985-
Dodd, Mrs H.P., C & P, 1985-
Hart, Mrs C.G., Mechanical, 1986-

*Retirement date unknown.

APPENDIX 2

Engineering School Enrolments

	Matriculated	Total
1890	5	53
1891	5	63
1892	8	62
1893	8	64
1894	6	59
1895	5	67
1896	5	86
1897	5	93
1898	11	98
1899	11	125
1900	11	127
1901	14	192
1902	39	190
1903	48	201
1904	52*	204
1905	45	177
1906	32*	160
1907	32	196
1908	29*	173
1909	26	178
1910	30*	181
1911	20	149
1912	22	133
1913	23	124
1914	24	124
1915	40*	107
1916	37	116

1917	32	107
1918	22	155
1919	31	187
1920	91*	269
1921	101*	257
1922	96*	250
1923	103*	234
1924	97*	238
1925	99*	250
1926	145	255
1927	149*	246
1928	158	266
1929	155*	249
1930	—	225
1931	—	220
1932	—	194
1933	—	171
1934	82	153
1935	75	153
1936	76	153
1937	74	209
1938	84	232
1939	—	201
1940	90	212
1941	99	192
1942	107	128
1943	97	169
1944	103	195
1945	218**	—
1946	460**	—
1947	418**	—
1948	349**	—
1949	285**	—
1950	257**	—
		Intermediates only
1951	284	84
1952	269	108
1953	256	102
1954	261	116
1955	276	133
1956	299	116
1957	325	117
1958	375	158
1959	422	167
1960	472	160
1961	513	180
1962	540	175
1963	559	175
1964	553	175
1965	618	78
1966	694	79

1967	677	83
1968	763	97
1969	796	114
1970	830	142
1971	829	164
1972	839	178
1973	852	156
1974	844	142
1975	841	138
1976	841	152
1977	809	148
1978	770	133
1979	821	145
1980	840	134
1981	839	127
1982	834	117
1983	831	117
1984	864	133
1985	895	159
1986	915	166

*Degree and Associate students
**Includes Intermediate students

APPENDIX 3

Deans of Engineering, 1923-1986

1924-28	Professor Percival Herbert Powell	Electrical
1929-32	Professor Samuel Steele	Mechanical
1933-34	Professor Powell	Electrical
1935-36	Professor Steele	Mechanical
1937-38	Professor Powell	Electrical
1939-41	Professor Steele	Mechanical
1942-43	Professor Powell	Electrical
1943-46	Professor Powell	Electrical
1946-49	Professor G.G. Calvert	Civil
1949-52	Professor R.J. Rastrick	Mechanical
1952-55	Professor N.M. MacElwee	Electrical
1955-58	Professor H.J. Hopkins	Civil
1958-61	Professor S.R. Siemon	Chemical
1961-64	Professor Rastrick	Mechanical
1964-67	Professor Hopkins	Civil
1967-70	Professor A. Miles Kennedy	Chemical
1970-74	Professor Leslie Kay*	Electrical
1974-77	Professor H. McCallion	Mechanical
1977-80	Professor Arthur G. Williamson	Chemical
1980-83	Professor David G. Elms	Civil
1983-86	Professor D. Clifford Stevenson	Mechanical
1986-	Dr Alexander J Sutherland	Civil

*Professor Stevenson was acting-Dean from early 1971 to August 1972 during Professor Kay's absence overseas.

APPENDIX 4

Numbers of Graduates

The numbers of degrees conferred in engineering at Canterbury (University) College and the University of Canterbury from 1896 to 1965 totalled 1933. The numbers since then are listed in greater detail below the first list.

Degrees in Engineering at Canterbury

Year	Number	Year	Number	Year	Number
1896	1	1921	4	1946	9
1897	1	1922	7	1947	26
1898	1	1923	8	1948	39
1899	—	1924	11	1949	64
1900	—	1925	11	1950	66
1901	—	1926	15	1951	73
1902	—	1927	15	1952	66
1903	—	1928	15	1953	64
1904	—	1929	25	1954	53
1905	3	1930	22	1955	66
1906	6	1931	20	1956	70 (7)
1907	4	1932	13	1957	57 (6)
1908	3	1933	27	1958	59 (2)
1909	6	1934	23	1959	75 (9)
1910	5	1935	21	1960	87 (9)
1911	4	1936	13	1961	75 (5)
1912	3	1937	20	1962	128 (7)
1913	—	1938	24	1963	127 (4)
1914	1	1939	19	1964	146 (3)
1915	6	1940	26	1965	122 (9)
1916	—	1941	14		....
1917	2	1942	14		1,933 (as at
1918	1	1943	10		May 1965)
1919	3	1944	19		
1920	2	1945	13		

Notes

(1) 15 of the first degrees conferred were B.Sc. degrees in Mechanical or Electrical Engineering; (the first graduate was G.A. Julius (Hon D.Sc. 1940))

(2) Additional figures in brackets indicate number of Masters Degrees in Engineering, (not included in totals for years 1956-65)

(3) Of all degrees conferred; 5% Chemical; 51% Civil; 30% Electrical and 14% Mechanical.

(4) A few cases of subsequent graduation in a second branch of Engineering are not included.

Numbers of engineering graduates since 1965 (with the numbers of women graduates included shown in brackets) have been as follows:—

Year	B.E.		M.E.	Ph.D.
1965	122	(1)	9	1
1966	124		7	4

1967	176	(1)	10		3	
1968	162		7		3	
1969	179	(3)	16		3	
1970	183	(3)	10		9	
1971	218	(2)	29	(2)	7	
1972	208		30		11	
1973	206	(1)	23		13	
1974	217	(1)	30		10	
1975	220	(2)	37	(1)	13	
1976	214	(4)	27		10	
1977	209	(3)	28	(1)	9	
1978	232	(6)	29		10	
1979	187	(3)	23	(2)	18	
1980	191	(9)	33		14	
1981	179	(4)	34	(1)	9	
1982	213	(9)	24		5	
1983	208	(6)	8		5	
1984	263	(10)	24	(1)	10	(1)
1985	216	(9)	18		10	
1986	205	(8)	31	(2)	11	

Certificates of Associate

In addition to degrees, 78 Certificates of Associate of the School were awarded between 1901 and 1933. The course for the Certificate was established in 1900 and was, according to *A History of Canterbury College* (p.62) . . . intermediate between purely Technical School instruction and the requirements for a University degree, except in the purely professional subjects in which the standard was the same as that of the degree.

APPENDIX 5

Postgraduate Enrolments

The numbers of postgraduate students in engineering since 1949 are shown below. Before 1955, students took a fourth year for an Honours B.E., which equates with the one-year M.E. students of 1955 and beyond.

	M.E.	Ph.D.	TOTAL
1949	(8)	1	(9)
1950	(8)	1	(9)
1951	(7)	2	(9)
1952	(12)	4	(16)
1953	(12)	4	(16)
1954	(11)	5	(16)
1955	15	4	19
1956	9	2	11
1957	7	2	9
1958	10	6	16
1959	5	9	14
1960	8	8	16
1961	7	13	20
1962	12	18	30

1963	9	26	35
1964	8	25	33

Numbers of Engineering Masters and Ph.D. candidates (including staff members) have been distributed between department as follows since 1965:

Year	ENCH			ENCI			ENEL			ENME		
	M.E.	Ph.D.	Total	M.E.	Ph.D.	Total	M.E.	Ph.D.	Total	M.E.	Ph.D.	Total
1965	2	6	8	2	15	17	6	10	16	2	3	5
1966	4	5	9	3	18	21	12	8	20	5	4	9
1967	3	6	9	2	20	22	13	11	24	5	6	11
1968	2	7	9	6	20	26	20	14	34	6	6	12
1969	3	6	9	12	23	35	16	21	37	6	6	12
1970	5	9	14	24	16	40	13	21	34	5	5	10
1971	2	11	13	19	19	38	15	24	39	3	8	11
1972	3	10	13	10	19	29	6	26	32	10	10	20
1973	6	8	14	15	17	32	21	25	46	4	11	15
1974	7	7	14	16	15	31	18	27	45	5	14	19
1975	8	11	19	14	12	26	11	22	33	9	22	31
1976	3	11	14	11	11	22	19	18	37	12	16	28
1977	2	9	11	19	12	31	14	22	36	6	18	24
1978	1	6	7	15	10	25	14	22	36	7	14	21
1979	4	7	11	18	8	26	16	23	39	4	11	15
1980	2	7	9	14	8	22	28	24	52	6	5	11
1981	3	5	8	13	7	20	18	21	39	4	6	10
1982	4	9	13	3	8	11	7	24	31	5	6	11
1983	3	9	12	15	9	24	11	22	33	5	7	12
1984	6	9	15	12	12	24	19	17	36	8	4	12
1985	6	8	14	22	14	36	20	21	41	4	5	9

Note: The increase in the proportion of Engineering students undertaking postgraduate studies has been:—

1950	2.9%
1960	3.4%
1970	11.9%
1980	11,2%
1985	11.2%

APPENDIX 6

Graduate Distribution by Departments

Between branches of engineering, graduates were distributed as follows:

B.E. degrees	ENCH %	ENCI %	ENEL %	ENME %
1965	14	39	33	14
1970	16	44	25	15
1975	13	38	24	25
1980	8	38	37	17
1985	19	28	29	24

Postgraduate degrees	%	%	%	%
1965	10	40	40	10
1970	11	28	50	11
1975	10	42	38	10
1980	12	34	32	22
1985	—	43	43	14

APPENDIX 7

A Chronology of Buildings, 1890-1985

(Note: Prices, which are approximate, are expressed in $NZ).

1878-9	East wing, two lecture rooms, studies	B.W. Mountfort	$8,156
1890-1	Mechanical Engineering laboratory (adj.east wing, to e.)	B.W. Mountfort	$5,480
1901-2	Electrical Engineering laboratory (e.side n.quad.)	C.J. Mountfort	$6,906
1913-14	Hydraulics laboratory	Collins & Harman	$6,554
1920-3	Extension of School of Engineering	Collins & Harman	$29,740
1957-61	New School of Engineering at Ilam	M.O.W.	$1,316,646
1963-9	School of Engineering Extensions	M.O.W.	
1963-4	Stage 1 Chemical		$55,174
1964-5	Stage 2 Mechanical		$108,398
1964-7	Stage 3 Civil & Electrical		$246,152
1966-7	Stage 4 Lecture Theatres		$193,124
1967-9	Stage 5 Chemical Main Extension		$328,481
1965-7	Stage 6 Lecture Theatres & Library		$339,984
1968-70	Stage 7 Internal Alterations		$100,000
1967-77	Draughting Room Mechanical	F.R. Smith	$15,000
1984-85	Library extension	M.O.W.	$598,000

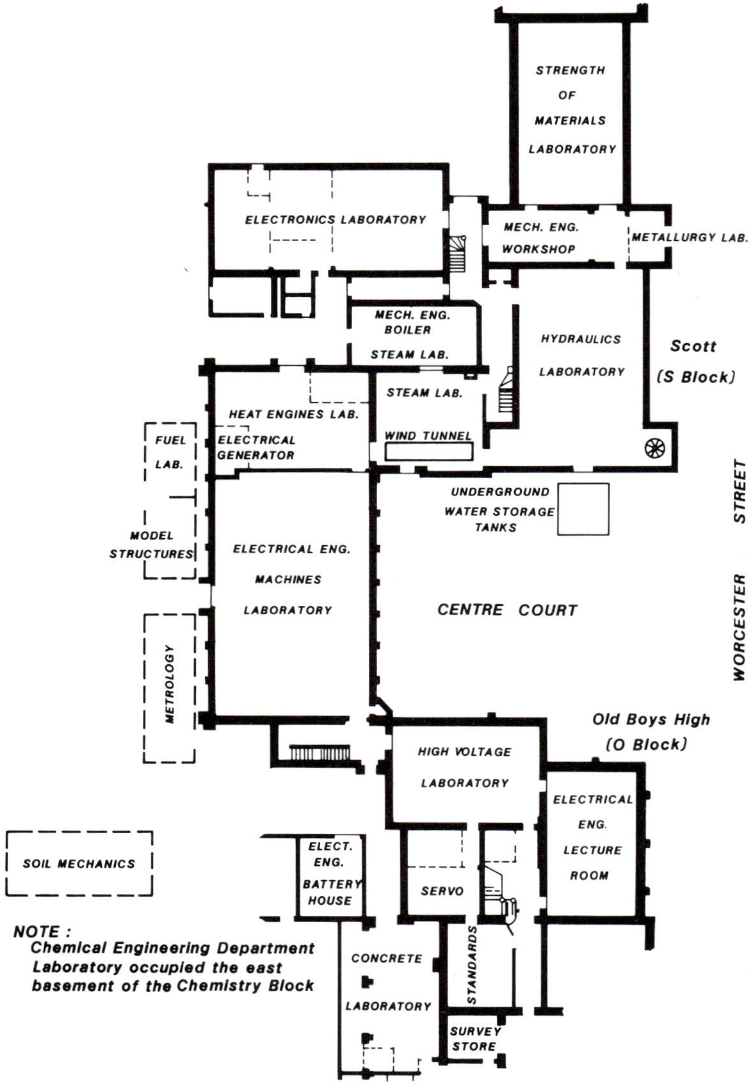

NOTE :
Chemical Engineering Department Laboratory occupied the east basement of the Chemistry Block

THE NATIONAL
SCHOOL OF ENGINEERING
CANTERBURY UNIVERSITY COLLEGE
LEVEL 1 - 1957

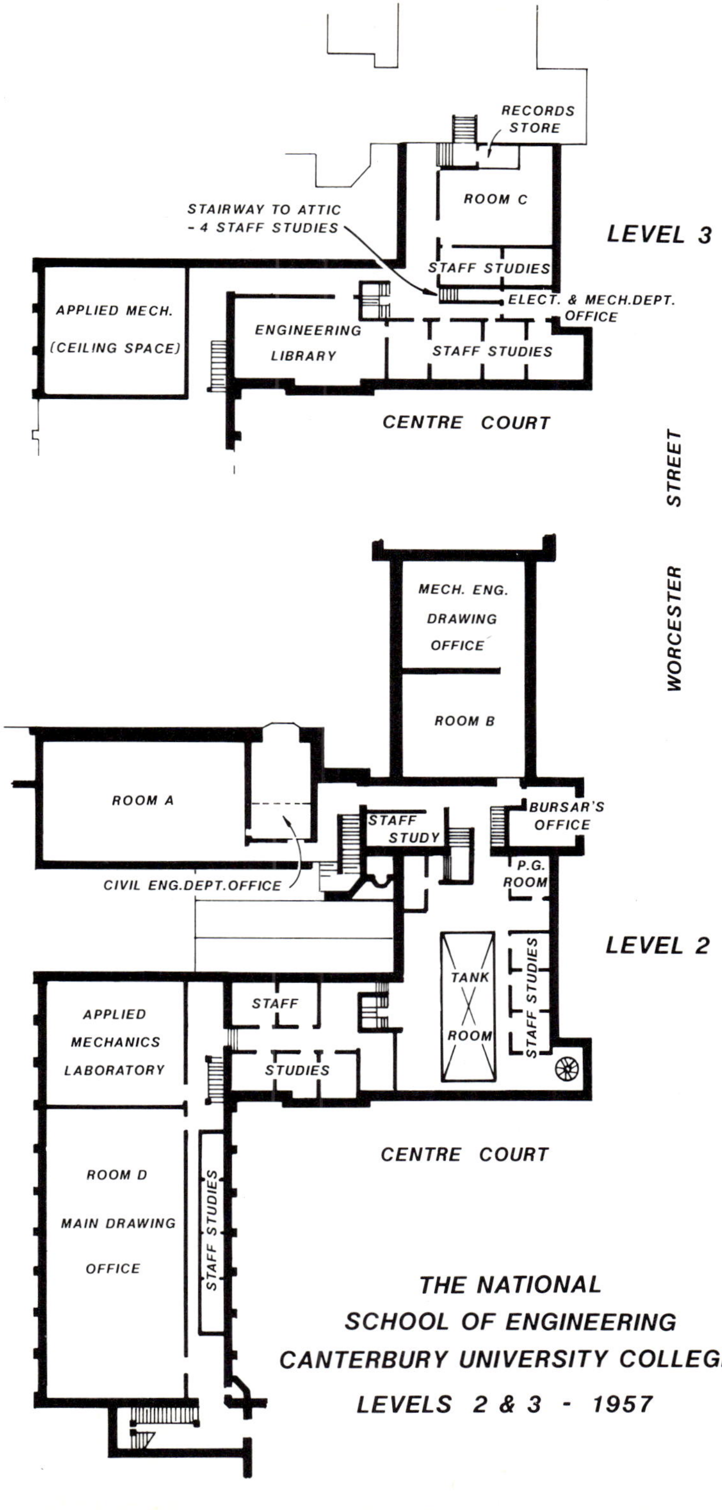

THE NATIONAL
SCHOOL OF ENGINEERING
CANTERBURY UNIVERSITY COLLEGE
LEVELS 2 & 3 - 1957

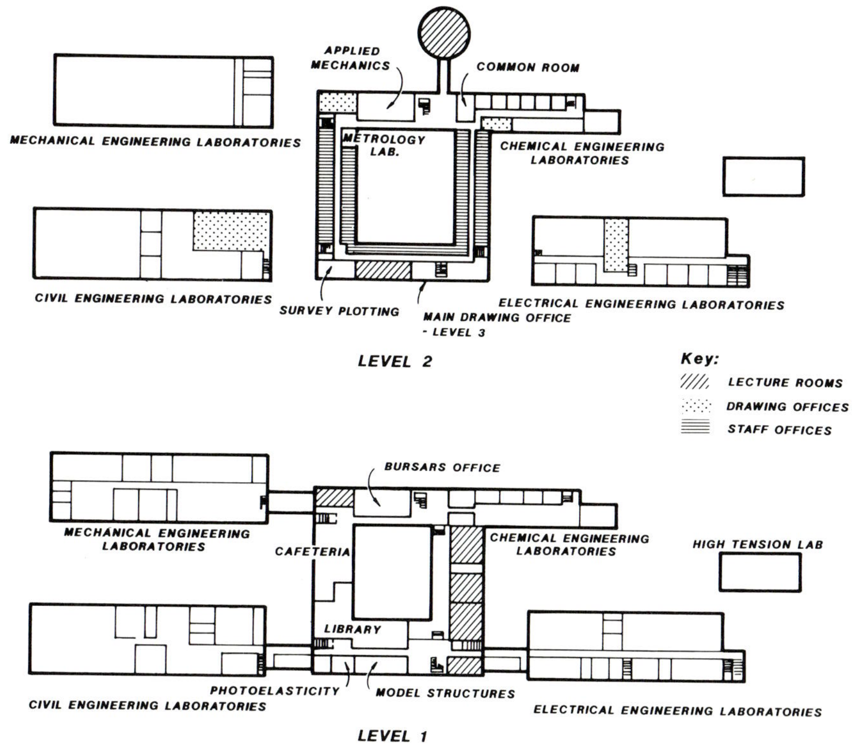

SCHOOL OF ENGINEERING UNIVERSITY OF CANTERBURY - 1961

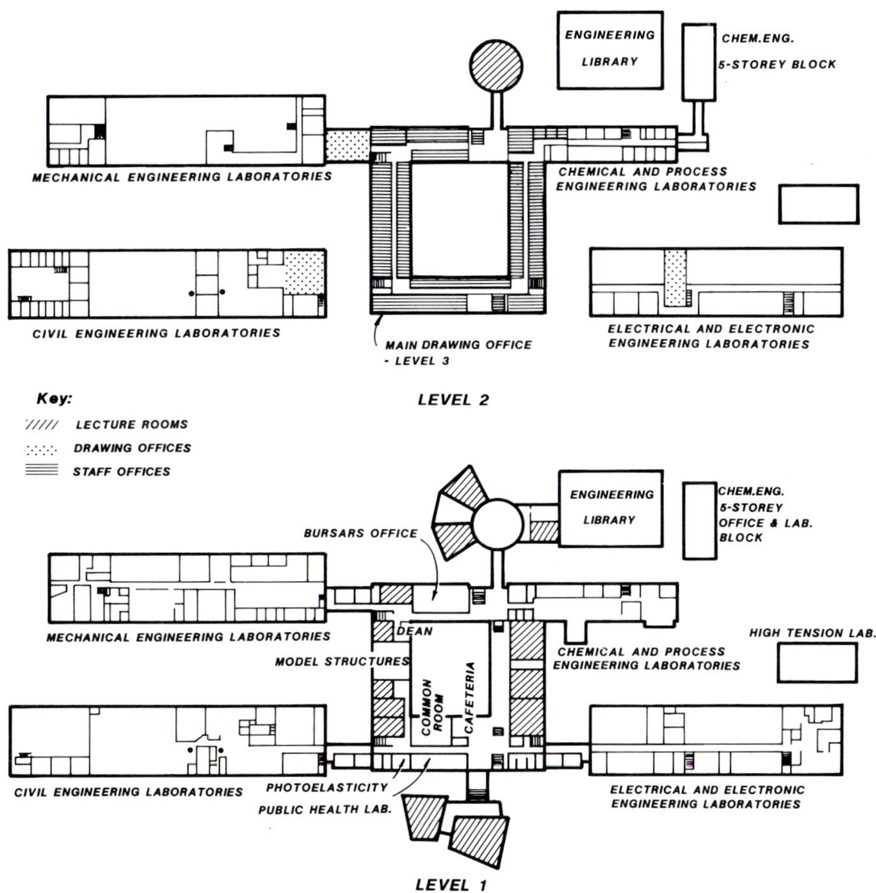

SCHOOL OF ENGINEERING UNIVERSITY OF CANTERBURY - 1987

Notes On Sources

PART ONE
Engineering Potentate

Chapter 1. *Introduction: 1849-1889.*

[1] *LT*, 29 March 1887.
[2] *P*, 22 Feb 1887.
[3] quoted in James Hight and C.R.Straubel (eds), *A History of Canterbury*, vol. 1, p.179.
[4] A.H.Reed, (ed.), *With Anthony Trollope in New Zealand*, p.59.
[5] W.J.Oliver, *The Story of New Zealand*, p.18.
[6] *opp. cit.*, p.65.
[7] *Canterbury Papers*, p.98.
[8] G.E.Emmerson, *Engineering Education: A Social History*, p.119.
[9] *A History of the University of Canterbury*, p.52.
[10] H.J.Tancred, "Inaugural Address", p.13.
[11] *LT*, 11 March 1879.
[12] *LT*, 9 Feb 1877.
[13] *LT*, 10 July 1882.
[14] *LT*, 29 March 1887.
[15] *LT*, 27 April 1887.
[16] *LT*, 26 March 1884.
[17] *LT*, 24 June 1884.
[18] C.W.Purnell, "A System of Technical Education for Artizans", *Transactions, Proceedings of New Zealand Institute*, 3 April, 1884, p.398.
[19] Sir Robert Stout, "Public Education in New Zealand", 18 June 1886, p.13.
[20] *LT*, 28 March 1887.
[21] *P*, 20 Oct 1941.
[22] *A History of the University of Canterbury*, p.59.
[23] *LT*, 5 April 1887.
[24] *LT*, 27 April 1887.
[25] *LT*, 13 April 1887.
[26] *LT*, 12 July 1888.
[27] *A History of the University of Canterbury*, p.132.
[28] Heinrich von Haast, *The Life and Times of Sir Julius von Haast*, pp.167-8.
[29] Arnold Wall, *Long and Happy*, p.107.
[30] G.R.Macdonald, "Canterbury Biographical Dictionary", on R.J.Scott.
[31] *P*, 5 Nov 1928.
[32] R.J.Scott, Unpublished typescript, pp.4-5.
[33] *P*, 10 Nov 1930.
[34] LO, 1 July 1909.
[35] *CCR*, Oct 1900, p.7.

[36] *P*, 10 Nov 1930.
[37] Personal reminiscence.
[38] *A History of the University of Canterbury*, p.133.

Chapter 2. *School of Engineering: Humble Beginnings.*

[1] *LT*, 5 April 1887.
[2] LI, 10 May 1897.
[3] *LT*, 25 March 1890.
[4] *P*, 10 Nov 1930.
[5] *A History of the University of Canterbury*, p.51.
[6] AR, 1893, p.9.
[7] F.W.Furkert, *Early New Zealand Engineers*, p.263.
[8] *LT*, 15 July 1890.
[9] LI, 1 Feb 1891.
[10] LI, 10 Dec 1891.
[11] A.P.W.Kennedy, *Minutes of the Proceedings of the Institution of Civil Engineers*, vol LXXXVIII, pp.1-2.
[12] AR, 1892, p.4.
[13] AR, 1893, p.10.
[14] Ibid, p.10.
[15] LI, 9 Aug 1893.
[16] *P*, 4 June 1894.
[17] *P*, 8 June 1894.
[18] LO, Sept 1893.
[19] LI, 9 June 1894.
[20] LI, 4 July 1894.
[21] LI, 4 July 1894.
[22] LI, 28 Sept 1894.
[23] LI, 5 Feb 1895.
[24] *P*, 4 June 1894.
[25] *LT*, 10 Sept 1894.
[26] *LT*, 5 Sept 1895.
[27] *LT*, 10 Sept 1895.
[28] LI, 31 July 1895.
[29] *LT*, 31 Oct 1889.
[30] AR, 1895, p.8.

Chapter 3. *The First Decade.*

[1] LO, 22 Nov 1892.
[2] LI, 26 Nov 1892.
[3] James Hight and Alice Candy, *A Short History of Canterbury College*, p.59.
[4] LI, 3 March 1893.
[5] *P*, 4 June 1892.
[6] LO, April, 1897.
[7] LO, March 1897.
[8] LI, 10 May 1897.
[9] LI, 29 May 1897.
[10] LI, 21 Dec 1889.
[11] LI, 28 July 1896.
[12] LI, 1 March 1897.
[13] LI, 5 March 1897.

[14] LI, 20 Feb 1899.
[15] LI, 1 March 1899.
[16] Arnold Wall, *Long and Happy*, p. 107.
[17] LO, March 1899.
[18] *P*, 3 Feb 1900.

Chapter 4. "*A Separate Institution.*"

[1] *LT*, 1 March 1897.
[2] *P*, 4 June 1894.
[3] *Canta*, 2 July 1937.
[4] *P*, 24 June 1899.
[5] LI, 20 July 1903.
[6] *P*, 26 June 1906.
[7] *Ibid.*
[8] LI, 11 June 1906.
[9] LO, 26 June 1906.
[10] *P*, 24 June 1899.
[11] *CCR*, 1900, p.34.

Chapter 5. *Electrical Laboratory.*

[1] LI, 27 March 1907.
[2] AR, 1900, p.10.
[3] Roslyn Noonan, *By Design*, p.90.
[4] *P*, 14 June 1900.
[5] *P*, 12 Jan 1901.
[6] LI, 7 Jan 1901.
[7] AR, 1901, p.13.
[8] *P*, 28 June 1902.
[9] *P*, 12 Jan 1901.
[10] *CCR*, June 1902, p.14.
[11] *CCR*, June 1903, p.29.
[12] *P*, 12 Jan 1901.
[13] *P*, 25 March 1901.
[14] *P*, 23 Jan 1902.
[15] *CCR*, June 1902, p.2.
[16] *P*, 28 June 1902.
[17] *Ibid.*
[18] *Ibid.*
[19] *Ibid.*
[20] *Ibid.*
[21] *Ibid.*
[22] LI, 17 July 1903.
[23] LI, 1 July 1903.
[24] LI, 20 July 1903.
[25] Ibid.
[26] Ibid.
[27] LI, 19 Aug 1904.
[28] LI, 19 Jan 1904.
[29] *NZ Engineering*, vol.14, no 2, 15 Feb 1959.
[30] LI, 2 Feb 1911.

Chapter 6. *Civil Engineering.*

[1] LI, 28 Aug 1903.
[2] LI, 24 March 1905.
[3] LI, R.O'Shea, Registrar, AUC, 1931.
[4] *P*, 25 July 1905.
[5] *LT*, 28 June 1905.
[6] *P*, 22 Feb 1907.
[7] *P*, 18 June 1910.
[8] LI, 5 July 1910.
[9] LI, 15 Oct 1912.
[10] *P*, 30 April 1912.
[11] LI, 9 Oct 1912.
[12] LI, 15 Oct 1912.
[13] *A History of the University of Canterbury*, p.186.
[14] LI, 29 Jan 1913.
[15] *P*, 31 Jan 1913.
[16] LI, 18 Feb 1914.
[17] LI, 3 June, 1914.
[18] LI, 3 March 1913.
[19] *P*, 29 April 1914.
[20] *CCR*, Oct 1897, p.1.
[21] *Ibid*, p.17.
[22] *CCR*, Oct 1898, p.24.
[23] *LT*, 1 April 1903.
[24] *P*, 15 Feb 1924.
[25] AR, 1907, p.3.
[26] *P*, 26 March 1907.
[27] LI, 8 Feb 1906.

Chapter 7. *Room for Infinite Ructions.*

[1] *P*, 22 Nov 1908.
[2] *P*, 11 Dec 1908.
[3] *Ibid.*
[4] *P*, 22 Nov 1908.
[5] *P*, 3 Nov 1925.
[6] *P*, 22 Nov 1908.
[7] *A History of the University of Canterbury*, p.135.
[8] LI, July 1910.
[9] *P*, 13 June 1908.
[10] LI, 13 Nov 1909.
[11] LI, 14 Nov 1910.
[12] LI, 5 July 1910.

Chapter 8. *War: 1914-1918.*

[1] AR, 1915, p.23.
[2] AR, 1917, P.30.
[3] LI, 28 April 1914.
[4] LI, 6 Aug 1914.
[5] *P*, 28 March 1916.
[6] AR, 1919, p.29.
[7] MLC, 18 June 1919.

[8] LO, 17 Sept 1919.
[9] LO, 16 May 1919.
[10] *P*, 28 June 1921.
[11] LI, 22 Sept 1905.
[12] LO, 24 Nov 1905.
[13] LI, 2 Feb 1906.

Chapter 9. *Scott's Retirement.*

[1] *P*, 5 May 1923.
[2] LI, 5 Aug 1921.
[3] *P*, 28 Sept 1922.
[4] *P*, 9 April 1923.
[5] LI, 24 Feb 1922.
[6] Personal Reminiscence.
[7] LI, 26 Feb 1913.
[8] *P*, 6 Sept 1907.
[9] *College Rhymes*, "Two Miniatures", 1919.
[10] LI, 7 June 1921.
[11] LI, 15 Feb 1932.
[12] LI, 24 Feb 1922.
[13] LI, 3 May 1922.
[14] *P*, 9 April, 1923.
[15] *P*, 28 Sept 1922.
[16] LI, 3 May 1922.
[17] *A History of the University of Canterbury*, p.199.
[18] LI, 7 June 1922.
[19] LI, 20 Jan 1923.
[20] *CCR*, June 1903, p.28.
[21] *P*, 27 Jan 1920.
[22] *P*, 10 Nov 1930.
[23] *P*, 12 Nov 1930.
[24] *P*, 10 Nov 1930.
[25] *Ibid.*

Chapter 10. *Organisation of the School: A Director or a Faculty.*

[1] CC, 20 Oct 1924.
[2] LI, 23 Feb 1955.
[3] LO, 5 June 1924.
[4] LI, 19 June 1924.
[5] *P*, 20 March 1929.
[6] *P*, 31 Oct 1928.
[7] *P*, 31 Oct 1928.
[8] *P*, 5 Nov 1928.
[9] *P*, 31 Oct 1928.
[10] *P*, 30 May 1927.
[11] LI, 16 Oct 1928.
[12] LI, April 1928.
[13] LI, 7 June 1921.

PART TWO
In The Doldrums

Chapter 1. *At a Standstill.*

[1] *P*, 5 Feb 1926.
[2] LI, 31 May 1926.
[3] LI, 19 Oct 1929.
[4] LI, 7 Nov 1929.
[5] LI, 18 Nov 1929.
[6] LI, 7 Nov 1929.
[7] Overseas examiners' report, 1924, p.17.
[8] Ibid, p.17.
[9] Overseas examiners' report, 1926, p.15.
[10] Overseas examiners' report, 1927, p.20.
[11] Overseas examiners' report, 1924, p.18.
[12] Overseas examiners' report, 1925, p.13.
[13] Overseas examiners' report, 1926, p.21.
[14] Overseas examiners' report, 1924, p.18.
[15] LI, 27 Sept 1923.

Chapter 2. *The School of Engineering: A "Rabbit Warren".*

[1] LI, 8 Nov 1941.
[2] *P*, 3 Nov 1959.
[3] *Canta*, 15 July 1932, p.1.
[4] *Canta*, 5 Aug 1932, p.3.
[5] LI, 20 Dec 1944.
[6] LI, 31 Aug 1943.
[7] LI, 19 Jan 1943.
[8] *P*, 18 June 1910.
[9] *A History of the University of Canterbury*, p.347.
[10] LI, 20 Dec 1944.
[11] LI, 10 Sept 1945.
[12] LI, 6 March 1946.
[13] LI, 8 July 1948.
[14] LI, 22 Sept 1944.
[15] LI, 27 March 1947.
[16] John Milton, "Il Penseroso".
[17] LI, 15 Sept 1948.
[18] *P*, 18 Sept 1952.

Chapter 3. *Staffing Difficulties: Professors, 1923-59.*

[1] LI, 22 April 1922.
[2] LI, 5 April 1946.
[3] LI, 19 Oct 1934.
[4] LI, 14 June 1945.
[5] LI, 25 Jan 1945.
[6] LI, 18 Feb 1951.
[7] LI, 23 Feb 1951.
[8] LO, 21 March 1951.
[9] LO, 29 March 1951.

[10] LO, 13 May 1953.
[11] *Canta*, 11 June 1953.
[12] LI, 25 Jan 1945.
[13] LI, 9 July 1945.

Chapter 4. *Staffing Difficulties: Junior Staff, 1923-59.*

[1] LI, 11 Sept 1944.
[2] LI, 31 July 1935.
[3] LI, 8 April 1936.
[4] *P*, 20 March 1929.
[5] LI, 12 Oct 1936.
[6] LI, 18 Nov 1936.
[7] LI, 27 May 1944.
[8] LI, 17 Oct 1946.
[9] LI, 9 Sept 1949.
[10] LI, 23 July 1949.
[11] LI, 26 June 1947.
[12] LI, 9 June 1950.
[13] *A History of the University of Canterbury*, pp.361-2.
[14] LI, 16 Sept 1939.
[15] LI, 20 Sept 1945.
[16] LI, 28 Nov 1922.
[17] LI, 15 May 1922.

Chapter 5. *The Depression.*

[1] *A History of the University of Canterbury*, p.180.
[2] *P*, 5 March 1921.
[3] LI, 20 Sept 1927.
[4] LI, 19 Oct 1934.
[5] LI, 12 Feb 1931.
[6] LI, 6 Sept 1933.
[7] LO, 20 Feb 1931.
[8] LI, 6 Oct 1934.
[9] LI, 14 Aug 1932.
[10] AR, 1936, p.23.

Chapter 6. *Provincial Rivalry.*

[1] J.C.Beaglehole, *The University of New Zealand: an Historical Study*, p.226.
[2] LI, 27 Feb 1931.
[3] *A History of the University of Canterbury*, p.134.
[4] Judith Bassett, *The School of Engineering, University of Auckland*, p.27.

Chapter 7. *Canta.*

[1] *Canta*, 2 Oct 1941.
[2] *P*, 28 Oct 1941.
[3] *Ibid.*
[4] *Canta*, 5 Nov 1942.
[5] *Ibid.*
[6] LI, 17 Jan 1943.
[7] Personal Reminiscence.
[8] LI, 19 Jan 1943.

[9] LI, 5 Nov 1941.
[10] Personal Reminiscence.
[11] Personal Reminiscence.
[12] Personal Reminisence.
[13] LI, 19 Jan 1943.
[14] AR, 1930, p.27.
[15] LI, 22 Oct 1936.
[16] LI, 17 May 1943.
[17] LO, 2 April 1946.
[18] LI, 17 Aug 1944.
[19] LO, 2 April 1946.

Chapter 8. *Civil Engineering.*

[1] LI, 9 Aug 1937.
[2] LI, 27 Sept 1937.
[3] LI, 20 March 1939.
[4] LI, 16 June 1944.
[5] LI, 6 Sept 1943.
[6] LI, 25 June 1945.

Chapter 9. *Chemical Engineering.*

[1] quoted in LI, 20 March 1943.
[2] LI, 13 Oct 1937.
[3] LO, 22 April 1938.
[4] LI, 18 May 1937.
[5] LI, 16 March 1943.
[6] LI, 26 March 1943.
[7] LI, 20 Sept 1945.
[8] LO, 22 Feb 1944.
[9] LI, 3 Oct 1945.

Chapter 10. *Electronics.*

[1] LI, 2 March 1939.
[2] LI, 7 Oct 1939.
[3] LI, 13 Nov 1939.
[4] LI, 19 Nov 1942.
[5] LI, 16 Dec 1942.
[6] LI, 17 May 1943.
[7] LO, 14 June 1943.
[8] LI, 27 June 1921.
[9] LI, Jan 1948.
[10] LI, 29 Aug 1945.
[11] LI, 27 March 1946.
[12] LI, 4 April, 1946.
[13] LI, 14 June 1947.
[14] LI, 15 Sept 1947.
[15] LI, 20 June 1946.
[16] LI, Aug 1947.
[17] LO, Aug 1947.
[18] LI, 13 Nov 1947.
[19] LI 27 Feb 1948.

[20] LI, 17 April 1948.
[21] LI, 18 Sept 1948.
[22] LI, 20 Sept 1948.
[23] Sept 1948 Report of Special Committee re Professor MacElwee and Mr Pollard.
[24] LI, 13 June 1949.
[25] LI, 21 July, 1949.
[26] LI, 12 Oct 1948.
[27] LI, 28 Nov 1958.

Chapter 11. *Breaking-Point.*

[1] LI, 25 Aug 1944.
[2] LO, 22 Dec 1944.
[3] LI, 5 Feb 1945.
[4] *P*, 1 May 1947.
[5] LI, 22 Sept 1947.
[6] Engineering Sub-Committee, 12 Dec 1948.
[7] *P*, 8 July 1949.
[8] *A History of the University of Canterbury, p. 336.*
[9] LI, 1 Dec 1949.
[10] LI, 1 Dec 1949.
[11] LI, 9 Dec 1949.
[12] *Star*, 8 July 1947.
[13] CM, 26 June 1950, p.35.
[14] LO, 15 June 1950.
[15] LO, 16 Feb 1951.
[16] LI, 16 Feb 1951.
[17] LI, 13 April 1951.
[18] LI, June 1951.

Chapter 12. *The Long Wait.*

[1] LO, 16 Sept 1949.
[2] LI, 16 June 1950.
[3] J.C.Beaglehole, *The University of New Zealand: an Historical Study*, pp.388-89.
[4] 1954 Report to the Senate, p.4.
[5] LI, 26 July 1949.
[6] *Star*, 31 Aug 1953.
[7] LI, 18 March 1952.
[8] LI, 13 March 1952.
[9] LI, 17 March 1952.
[10] LI, 18 March 1955.
[11] Personal Reminiscence.
[12] LI, 25 Aug 1944.
[13] LI, 14 Feb 1923.
[14] *P*, 3 Nov 1959.

PART THREE
A Terrible Beauty

Chapter 1. *The Long Haut To Ilam*

[1] *P*, 10 May 1958.
[2] N.C. Phillips, *A History of the University of Canterbury 1873-1973* p.443.
[3] Letter to Engg. Ctee., UR file 4958, 10 Feb 1952.
[4] *P* 27 May 1947.
[5] *SS* 8 July 1947.
[6] CM 17 Dec. 1948.
[7] Harman to Ward, UR file 4648, 20 Jan 1949.
[8] *Un. History*, p335.
[9] CM 25 Feb. 1949.
[10] CM 21 March 1949.
[11] Ibid.
[12] *P* 4 June 1949.
[13] Ibid 4 July 1949.
[14] Ibid 8 July 1949.
[15] Ibid 27 Aug 1949.
[16] PBM 23 Feb 1950.
[17] CM 28 Feb 1950.
[18] *P* 30 Nov 1950.
[19] Beeby to Registar, UR file 4704, 2 May 1951.
[20] "Some Notes on the Proposed Plan for new University Buildings at Riccarton" by R.J. Rastrick, UR file 4958, 16 Aug 1951.
[21] Ibid.
[22] *P* 29 March 1952.
[23] UR file 4958, 1950.
[24] Hulme to Dean, UR file 4958, 6 June 1951.
[25] Institute to Registrar, UR file 4958, 19 Oct 1951.
[26] Harman to Registrar, ibid, 26 Oct 1951.
[27] Logie to Harman, ibid, 6 Nov 1951.
[28] *P* 13 Sept 1951.
[29] *University History*, p324.
[30] Beeby to Rector, UR file 4958, 7 Feb 1952.
[31] *P* 26 Feb 1952.
[32] Ibid.
[33] *P* 27 Feb 1952.
[34] *SS* 26 Feb 1952.
[35] Institution to Registrar, UR file 4958, 10 March 1952.
[36] *P* 3 Aug 1951.
[37] Ibid 27 Feb 1952.
[38] Staff to Rector, UR file 4958, 26 Feb 1952.
[39] Staff to Rector, ibid, 5 March 1952.
[40] *P* 21 March 1952.
[41] Ibid 26 March 1952.
[42] *SS* 21 March 1952.
[43] Mins of Meeting, UR File 4958/1.10 March 1952.
[44] *P* 29 March 1952.
[45] Ibid 1 April 1952.

[46] Ibid 2 April 1952.
[47] Logie to Beeby, UR file 4958, 28 April 1952.
[48] *P* 21 May 1952.
[49] *SS* 21 May 1952.
[50] Engineering Advisory Ctee Report, UR file 4958/1,14 May 1952
[51] Mins of mtg, UR file 4958, 28 April 1952.
[52] *P* 10 June 1952.
[53] Ibid.
[54] Ibid, 11 June 1952.
[55] *SS* 10 June 1952.
[56] CM 30 June 1952.
[57] *P* 29 Oct 1952.
[58] Ibid 7 Nov 1952.
[59] Ibid 25 Nov 1952.
[60] *SS* 5 Nov 1952.
[61] Ibid 23 Feb 1953.
[62] *P* 24 Feb 1953.
[63] *New Zealand Engineering,* 15 April 1953.
[64] Logie to Algie, UR file 4958, 29 July 1953.
[65] *SS* 12 June 1953.
[66] *P* 1 July 1953
[67] Ibid 1 Sept 1953.
[68] N.Z. Parliamentary Debates, Vol 301, pp2134-39.
[69] *SS* 5 Nov 1953.
[70] *P* 28 Nov 1953.
[71] Ibid 8 May 1954.
[72] Algie to Cartwright, UR file 4958, 24 May 1954.
[73] Algie to Bain, UR file 4958, 24 Aug 1954.
[74] *P* 5 Nov 1957.
[75] Sub-ctee Mins, UR file 4958, 15 Oct 1954.
[76] *P* 29 March 1955.
[77] University History, p367.
[78] Rector to Govt Architect, UR file 4958, 27 April 1956.
[79] Rector to Govt Architect, ibid, 26 June 1956.
[80] Govt Architect to Rector, ibid, 26 July 1956.
[81] *P* 28 Aug 1956.
[82] Hopkins to Rector, UR file 4958, 11 Sept 1956.
[83] *P* 19 Sept 1956.
[84] Rector to Prime Minister, UR file 4958, 21 Sept 1956.
[85] *P* 13 Dec 1956.
[86] Un. *Chronicle*, Vol 12 No.23, Dec 1977.
[87] Hanson to Beeby, UR file 4958, 4 June 1957.
[88] *P* 29 Oct 1957.
[89] CM 29 Oct 1957.
[90] *P* 3 Dec 1957.
[91] Hopkins to Rector, UR file 4958A, 13 May 1959.
[92] Rector to Hopkins, Ibid, 21 May 1959.
[93] Hopkins to Rector, Ibid, 4 June 1959.
[94] CM 21 Sept 1959.
[95] Rector to Beeby, UR file 4958/1, 19 Oct 1959.
[96] M.F. Miles to Govt. Architect, UR file 4958, 13 July 1960.
[97] J.H. Hutchinson Education Dept to Registrar, UR file 4958E, 15 May 1969.

[98] Hopkins to Vice-Chancellor, UR file 4958, 25 Jan 1961.
[99] Currie to Llewellyn, UR file 5063, 17 Sept 1960.
[100] V.C. to UGC, UR file 4958/1, 11 June 1962.
[101] V.C. to Llewellyn, ibid, 5 Nov 1962.
[102] Ibid.
[103] *P* 16 Oct 1962.
[104] Harman to Registrar, UR file 4958, 16 Nov 1951.
[105] *P* 5 Nov 1957.
[106] Ibid, 14 Aug 1965.
[107] Ibid 24 April 1963

Chapter 2. *A Library Of Substance*

[1] Dean to Librarian, Library file, June 1943.
[2] "Memo on Requirements of the Engineering Departmental Library", UR file 4958/1, Aug 1951.
[3] "Report on the Proposed School of Engineering at Riccarton", UR file 4958, Sept 1951.
[4] Rector on Engineering School specifications, UR file 4958/1, 1 Oct 1951.
[5] Collins to Einhorn (MoW), Library file, 17 May 1955.
[6] Ibid.
[7] Siemon to Llewellyn, UR file 4958, 4 March 1960.
[8] Pownall to Tennent, UR file 4958, 12 June 1962.
[9] Rastrick to N.C. Phillips, Library file, 13 March 1964.
[10] Collins to Library Ctee, UR file 4968, 13 March 1967.
[11] Ibid.
[12] V.C. to Library Ctee, UR file 4958, 31 March 1967.
[13] Collins to Bursar, Library file 20 June 1963.

Chapter 3. *College To University, National School To Faculty*

[1] *P*, 16 July 1958.
[2] MacElwee to Bain, UR file 5014, 26 Aug 1955.
[3] CM, 26 Sept 1955.
[4] *P* 16 Oct 1951.
[5] UR file 4801, 30 Nov 1949.
[6] Dean to Rector, UR file 4958/1, Sept 1951.
[7] Dean to Registrar, UR file 5014, 27 March 1975.
[8] Registrar to Dean, ibid, 8 April 1975.
[9] University *History* p 372.
[10] Rowley to V.C., UR file 5085, 11 Sept 1964.
[11] Engg. Fac. Mins, 15 Oct 1962.
[12] Planck to Pownall, UR file 5014, 19 Oct 1962.
[13] Pownall to Planck, ibid, 29 Oct 1962.
[14] Pownall to Dean, ibid, 26 Oct 1962.
[15] UC statistics.
[16] *P* 6 Nov 1964.
[17] UGC Report to Parliament, E3, 1980.
[18] Engg School quinquennial submission, UR file 5085, 1968.
[19] Report on quinquennial submission, ibid, 8 May 1968.
[20] *P* 5 Aug 1968.
[21] Ibid 8 Aug 1968.
[22] Eng. School quinquennial proposals, 8 Sept 1983.
[23] Ibid.

Chapter 4. *Elusive Expansion*

[1] John Hendry to UNZ, UR file 5220, 25 June, 1961.
[2] *P* 27 Aug 1960.
[3] Llewellyn to Pownall, UR file 5220, 16 Oct 1961.
[4] Hopkins to Pownall, ibid, 21 Sept 1965.
[5] Pownall to Llewellyn, ibid, 13 May 1965.
[6] Rowley to Pownall, ibid, 13 May 1966.
[7] Turbott to Llewellyn, ibid, 9 June 1966.
[8] N.C. Phillips, ibid, 7 May 1968.
[9] M.J. Moffat to V.C., ibid, 25 Sept 1969.
[10] Engg Fac Mins, 19 March 1980.
[11] *Engineering* (UGC Review Ctee,) Wgton 1982.
[12] V.C.'s report to Otago Un Cl, 18 June 1985.
[13] Canterbury submission, UR file 5014, Feb 1983.
[14] V.C.'s report to Otago Un Cl, 18 June 1985.
[15] Engg Fac Mins, Oct 1983.
[16] V.C. to J.D. Gould, UR file 5014, 2 Sept 1985.
[17] UGC working party report, UR file 5014, Oct 1983.
[18] Dean of Science submission, ibid, March 1984.
[19] Dean of Commerce submission, ibid, March 1984.
[20] Report to Engg Fac, ibid, 19 March 1984.
[21] *News VUW*, Vol 9, No 4, 6 April 1984.
[22] V.C. to UGC, UR file 5014, 7 Sept 1984.
[23] Ibid.
[24] "Performance of First-Year Engineering Students in Intermediate" - Report by U.G.C. ad hoc Sub-Ctee on Engg, Oct 1983.
[25] Ibid.
[26] Ibid.

Chapter 5. *Leading The School*

[1] Engg Fac mins, 9 Sept 1971.
[2] Ibid, 6 and 21 June 1972.
[3] Engg Fac Sub-ctee report, April 1986.
[4] Engg Fac Mins, 14 Aug 1985.
[5] PBM, 27 Aug 1986.
[6] N.C.P. in *The Times*, London, 20 Jan 1986.
[7] Un. *Chronicle*, Vol 12 No 23, 2 Dec 1977.
[8] Hopkins to chairman, 23 Sept 1952.
[9] Kay to V.C., UR file 5187, 13 May 1966.
[10] Nevill to V.C., UR file 5014, 28 Sept 1966.
[11] *P* 11 Sept 1967.
[12] Ibid 20 Aug 1969.
[13] Andreae to V.C., UR file 5187, 18 Feb 1971.
[14] Andreae to V.C., ibid,27 June 1971.
[15] V.C. to Andreae, ibid, 28 June 1971.
[16] Bates to V.C., UR file 5187e, 9 March 1973.
[17] Hudson Lecture to RSNZ, 26 April 1978.
[18] Bates to Chancellor, UR file 5187e, 17 Sept 1979.
[19] Un *Chronicle* Vol 11 No 20, 22 Oct 1976.
[20] *P* 24 Oct 1968.

Chapter 6. *A Research Revolution*

[1] *Karl Popper's Life and Scientific Work,* , Editrice L'Industria Milano Milan 1966, p 5.
[2] Nevill to UGC, UR file 5085, 21 Feb 1955.
[3] Hopkins to CUC Finance Ctee, ibid, 3 March 1955.
[4] Registrar to Hopkins, ibid, 14 March 1955.
[5] Un. *Chronicle* Vol 20 No7, 3 May 1985.
[6] *P* 30 June 1970.
[7] Un. *Chronicle* Vol 12 No 23, 2 Dec 1977.
[8] *P* 11 June 1968.
[9] *U.K. Sensor Review*, Jan 1985.
[10] *National Electronics Review*, U.K., July-Aug 1977.
[11] Robb to Bates, personal file, 16 June 1986.
[12] Un *Chronicle*,Vol 8 No 2, July 1973.
[13] Ibid, Vol 21 No 3, 7 March 1986.
[14] Ibid.
[15] J.A. Ritchie to Byers and Harman, UR file 5187a, 14 July 1977.

Chapter 7. *Hard Work, Hard Play*

[1] Professor P.W. Taylor AU, Hopkins Lecture, 16 July 1980.
[2] Ibid.
[3] Dean of Science's submission to UGC, UR file 5014, 20 March 1984.
[4] Ibid.
[5] *P* 2 Nov 1981.
[6] *P* 3 Nov 1981.
[7] Women graduates to author, personal file, 1986.
[8] *P* 14 Sept 1983.
[9] Diploma Day song, 1901.
[10] *Student Engineer* Vol 111, March 1950.
[11] Ensoc annual report 1959.
[12] Ibid, 1966
[13] Ibid.
[14] Ensoc social committee annual report, 1967.
[15] Ensoc annual report 1948.
[16] Gordon Slatter, *The Story of Rolleston House*, p.109.
[17] *Student Engineer* Vol 20, Nov 1967.
[18] Ibid, Vol 16, Nov 1963.
[19] Ensoc annual report, 1963.
[20] *P* 12 June 1964.
[21] *Student Engineer* Vol 17, Nov 1964.
[22] *P* 13 Nov 1971.
[23] Ibid 3 Oct 1969.
[24] *Student Engineer* Vol 12 Nov 1959.
[25] Ibid, Vol 21, Nov 1968.
[26] *P* 28 April 1978.
[27] Ibid 15 Aug 1978.
[28] Un. *Chronicle* Vol 17, No.20, 29 Oct 1982.

Chapter 8. *Retrospect And Prospect*

[1] Un. *Chronicle* Vol 11, No. 15, 3 Aug 1976.
[2] *P* 12 Sept 1962.
[3] Ibid, 4 April 1962.

[4] Ibid, 8 Dec 1964.
[5] Ibid, 30 July 1976.
[6] IPENZ education policy statement.
[7] C.M. Wheeler to V.C. UR file 5014, 19 Sept 1964.
[8] R.G. Norman, Hopkins Lecture, 16 July 1986.
[9] NE/NZ *Electronics Review* Sept 1985.
[10] *P* 7 June 1962.
[11] Fac Mins 19 March 1980.
[12] *P* 25 Nov 1966.
[13] C.G. Martin, Newnham Lecture, 1975.
[14] Horn to Registrar, UR file 4958. 8 July 1959.
[15] Bursar to Registrar, UR file 4958A, 7 July 1960.

Bibliography

Primary Sources

A. Manuscript or Unprinted

Records of the University of Canterbury, Canterbury University College and Canterbury College.

Minutes of the Board of Governors and Council of University and College
Minutes of Council and Board Committees
Minutes of the Professorial Board
Inward and outward correspondence of College and University.

B. Printed

(i) Records of University and College

Canterbury (University) College: Chairman's Annual Statement and Accounts, 1890-1948
Canterbury (University) College *Calendar* 1879-1957
University of Canterbury *Calendar* 1958-
University of Canterbury *Chronicle* 1963-

(ii) Records of the University of New Zealand and of the University Grants Committee.

Minutes of proceedings of the Senate, U.N.Z. 1875-1926 and 1929-61
Minutes of the Council, U.N.Z., 1927-28
Minutes of the Academic Board 1927-61
N.Z.U. *Calendar* 1873-1961
U.G.C. *Handbook* 1962-

(iii) N.Z. Government

Appendix to the Journals of the N.Z. House of Representatives (AJHR)
Papers relating to (Report on) the Canterbury College, 1878-1940
Relevant AJHR reports are:
1912, E 7a Finances of University and Colleges
1913, I 13a University Education in N.Z.
1925, E 7a *University Education* (Reichel-Tate Report)
1959 E 1, Problems of the University
1961- E3, Annual reports of the U.G.C. and University institutions
Report of the Committee on New Zealand Universities, 1959. (Hughes Parry report)
New Zealand Parliamentary Debates

(iv) Pamphlets, articles, lectures

Bates, R.H.T., Hudson Lecture, RSNZ 1978
Canterbury College (Students' Association) *Diploma Day Songs* 1890-1912. Macmillan Brown Collection

Chilton, Charles, *The Modern University, its Development and Functions*, Christchurch 1925.
Currie G.A. and Kedgley, E.G. *The University in New Zealand: Facts and Figures*, Wellington, 1959
Emery, T.G. *First Employment of University Graduates*, 1965-72
Hight J. *University Policies and Organisation*, Christchurch, 1928
Marsden, Sir Ernest, *The University and its Relations to Industry and the Community*, Christchurch 1931
Norman, R.G. Hopkins Lecture, Christchurch 1986
University of Canterbury Industrial Development Dept, *An Outline of the History and Operation of a New University Department*, Christchurch 1950

(v) Student publications
Canta, 1930-
Canterbury College Review, 1894-1948
Student Engineer, 1948-66

(vi) Newspapers and journals
Christchurch Star-Sun and *Christchurch Star*, Christchurch 1935-
Lyttelton Times and *Christchurch Times*, Lyttelton and Christchurch 1851-1935
The Press, Christchurch, 1861—
The Sun, Christchurch 1914-35
University of Canterbury newspaper clipping books
New Zealand Engineering, Wellington.

Secondary Sources

A Century of Achievement 1879-1979: A Record of Progress. Canterbury Manufacturers' Association, 1980
A History of Canterbury, Vol.11, 1854-76, W.J. Gardner (ed), Christchurch 1971; Vol.111, 1876-1950, W.J. Scotter, Christchurch 1965
A History of the University of Canterbury 1873-1973 by W.J. Gardner, E.T. Beardsley and T.E. Carter. (N.C. Phillips ed.), University of Canterbury, 1972.
Alpers, O.T.J. (ed) *College Rhymes: An Anthology of Verse*, Christchurch 1923.
Andersons Limited, Christchurch, 1850-1925, Printed by Whitcombe and Tombs Ltd, 1925
Armytage, W.H.G., *A Social History of Engineering*, Faber and Faber, London, 1961.
Ashby, Sir Eric, *Technology and the Academics*, Macmillan and Co. Ltd, London, 1958.
Bassett, Judith, *The School of Engineering, University of Auckland, 1906-1969: A History*, University of Auckland, 1969.
Beaglehole, J.C., *New Zealand: A Short History*, Allen and Unwin Ltd, London, 1936.
— *The University of New Zealand*, Wellington 1937
Buglianello, G. and Doner, D. (eds), *The History and Philosophy of Technology*, University of Illinois Press, Chicago, 1979.
Burdon, R.M., *The Life and Times of Sir Julius Vogel*, Caxton Press, Christchurch, 1948.
Burstall, Aubrey F., *A History of Mechanical Engineering*, Faber and Faber, London 1963.
Canterbury College Jubilee, 1873-1923, supplement to the *Lyttelton Times* Christchurch 1923.
Canterbury College 1873-1923; supplement to the *Press*, Christchurch 1923.
Christopherson, D.G., *Engineers and the University*, English Universities Press, Ltd, London, 1967.
Condliffe, J.B., *New Zealand in the Making: A Study of Economic and Social Development*, George Allen and Unwin, London, 1936.
Cross, Hardy, *Engineers and Ivory Towers*; New York, McGraw Hill 1952.
Social Development, George Allen and Unwin, London, 1936.
Dobson, Arthur Dudley, *Reminiscences*, Whitcombe and Tombs Ltd, New Zealand, 1930.

Dobson, Edward, *Pioneer Engineering*, Lockwood, London, 1880.
Dunsheath, Percy, *A History of Electrical Engineering*, London, Faber and Faber, 1962
Emmerson, George S., *Engineering Education: A Social History*, David and Charles: Newton and Abbot Crane, New York, 1973.
Florman, Samuel C., *The Existential Pleasures of Engineering*, St Martin's Press, New York, 1976.
— *Blaming Technology*, St Martin's Press, New York, 1979.
Furkert, F.W., *Early New Zealand Engineers*, A.H. and A.W. Reed, New Zealand, 1953
Gardner, W.J., *Colonial Cap and Gown*, University of Canterbury, No.25, 1979.
Gregory, Malcolm S., *The History and Development of Engineering*, Longman Group Ltd, London, 1971.
Hight J., Candy A., *A Short History of Canterbury College*, Whitcombe and Tombs, Christchurch 1927.
Hilken, T.J.N., *Engineering at Cambridge University, 1783-1965*, Cambridge University Press, 1967.
Hopkins, H.J., *A Span of Bridges*, David and Charles, Devon, 1970.
Keey, R.B. and others, *Engineering and Society*, University of Canterbury. No. 31, 1982.
Kirby, Richard Shelton, *Engineering in History*, New York, McGraw Hill 1956.
Morrell, W.P. and Hall, D.O.W. *A History of New Zealand Life*, Whitcombe and Tombs Ltd, New Zealand, 1957.
Mumford Lewis, *Technics and Civilisation*, George Routledge and Sons Ltd., London 1934.
Newnham, W.L., *Learning, Service, Achievement: Fifty Years of Engineering in New Zealand*, The New Zealand Institution of Engineers, 1971.
Noonan, Roslyn J., *By Design*, Crown Copyright, New Zealand 1975.
Oliver, W.J., *The Story of New Zealand*, Faber and Faber, London 1960.
Pannell, J.P.M., *An Illustrated History of Civil Engineering*, Thames and Hudson, London, 1964.
Parton, H.N., *The University of New Zealand*, University Grants Committee, 1979.
Reed, A.H. (ed), *With Anthony Trollope in New Zealand*, A.H. and A.W. Reed, New Zealand 1969.
Roller, Duane H.D. (ed), *Perspectives in the History of Science and Technology*, University of Oklahoma Press, 1971.
Simpson, Tony, *The Sugarbag Years: An Oral History of the 1930s Depression in New Zealand*, Alister Taylor, Wellington 1974.
Sinclair, Keith, *A History of New Zealand*, Penguin Books Ltd., Harmondsworth, Middlesex, England 1959.
Slatter, Gordon, *The Story of Rolleston House, 1919-74*, Pegasus Press, Christchurch 1977.
Straub, Hans, *A History of Civil Engineering; an outline from Ancient to Modern Times*; English translation by E.Rockwell; Leonard Hill, London, 1952.
University of Canterbury Industrial Development, *An Outline of the History and Operation of a new University Department*, Christchurch 1950.
Vogel Julius (ed), *The Official Handbook of New Zealand. A Collection of Papers by Experienced Colonists, on the Colony as a Whole and the Several Provinces*, Wyman and Sons, London, 1875.
von Haast, H.F. *The Life and Times of Sir Julius von Haast*, New Zealand, 1948.
Webb, Beatrice and Sidney, *Visit to New Zealand*, Milburn and Co, Wellington, 1959.

Index